# ZOOLOGY. ON (POST) MODERN ANIMALS

**1**

## artwerpen 93

*Culturele hoofdstad van Europa*
*Capitale culturelle de l'Europe*
*Kulturhauptstadt Europas*
*Cultural capital of Europe*
*Capital Europea de la cultura*

# ZOOLOGY. ON (POST) MODERN ANIMALS

*Edited by Bart Verschaffel & Mark Verminck*

THE LILLIPUT PRESS

This book is one in a series of six, realized as part of the project 'Vertoog & Literatuur' (Discourse and Literature) of Antwerpen 93, Cultural Capital of Europe. The six 'cahiers' (workbooks) are published in Dutch by the publishing house Kritak/Meulenhoff (Leuven/Amsterdam). The series consists of:
1. *Lijn, Grens, Horizon* (Line, Frontier, Horizon).
2. *Woordenloosheid* (Wordlessness).
3. *Provincialismen. Ontworteling* (Provincialisms. Uprootedness).
4. *Zoölogie. Over (post)moderne dieren* (Zoology. On (post)modern Animals).
5. *Orthodoxie (...) Applaus* (Orthodoxy [...] Applause).
6. *Restauraties. Over vormen van herstel* (Restorations. Forms of Repair).
Moreover, 'Cahiers' 1, 5 and 6 are published in French by Editions Mardaga (Liège/Belgium), 'Cahiers' 1 and 3 in German by Verlag J. Dinter (Köln/Germany) and 'Cahiers' 2 and 4 in English by The Lilliput Press (Dublin/Ireland).

This book is edited by Bart Verschaffel and Mark Verminck.

The editorial board consists of: Jan Denolf, Raymond Detrez, Geert Lernout, Rudi Laermans, Herman Parret, Patrick Vandermeersch and Marianne Van Kerkhoven.
Assistance: Ilse Vandingenen, Stefan Franck and Jeroen Olyslaegers.

*Acknowledgments:* Special thanks goes out to Kris Van De Poel. The text of Roger Avermaete was included and translated with kind permission of the Lions Club 'Bruxelles-Erasme'. Details of original publication are as follows: 'The Zoological Gardens' is a partial translation of 'Le jardin zoologique' in: Roger Avermaete, *Synthèse d'Anvers* (Antwerp 1932). 'Pense-bête' (Marcel Broodthaers) is a translation of a selection from the volume *Pense-Bête* (Brussels 1964). 'Cows' was taken from Witold Gombrowicz, *Diaries 1953-1969* (London 1986). 'The man with the swine's head' (Paul van Ostaijen) is a partial translation of 'De man met de zwijnekop' in *Verzameld Werk* (Amsterdam 1991).

Design by Antony Farrell. Layout by Jarlath Hayes. Cover design by Howard Noyes.
Formatted in 12 on 14 pt Monotype Bembo by STF, Celbridge. Printed in Dublin by Betaprint.
First published in 1993 by The Lilliput Press Ltd, 4 Rosemount Terrace, Arbour Hill, Dublin 7, Ireland.

ANTWERP 93 has settled the rights for the illustrations and the texts to the best of its ability. Further rightful claimants are invited to contact: ANTWERPEN 93; 'Vertoog en Literatuur', Grote Markt 29, B-2000 Antwerpen.

ISBN 1 874675 18 X

# CONTENTS

# EDITORS' PREFACE

Man has an ambiguous relationship with animals. He classifies and studies them, adopts them as pets, or rather, he anxiously keeps them outside, he lets them be of service or manipulates them, cuddles them, eats them ...

The place where all these ambiguous and contradictory attitudes are found at the same time, is undoubtedly the zoological garden: a place where the animal is caged *because* we love it, but also because it frightens us and fascinates us, amuses us and distinguishes us from them. The 'zoo', that nineteenth-century metropolitan area of town for animals, is at the same time Garden of Eden and Noah's ark, circus and amusement park, ghetto and jail ...

What is the place of the animals in our times? How do we relate to them — in the media, for instance? Has a new imagination with regard to the animals come into existence? Is the lion still 'courageous', the fox still 'sly', and the owl still 'wise' as the mythical animals in olden days? Are animals individuals or is that one of our inevitable projections?

The contributions to this book are about the zoo and modernity, the fable beasts from Aesop to cartoons, antique imagination about animals, 'natural histories', the horse's grandeur, bestiaries, designer chicken and swine fever, respect for animals.

This book or 'cahier' — a workbook — is published on the occasion of Ant-werpen 93, Cultural Capital of Europe, and is part of the Discourse and Literature programme. It offers a selection of original contributions by European novelists, poets, artists, essayists, literary critics, historians, sociologists, anthropologists, and philosophers. It also contains previously published materials and classic European texts.

*Zoology.On (post)modern animals* functions as a meeting-point and welcomes authors from numerous countries. Its purpose is to blend and merge into one the many genres which constitute Western written culture. Poetry and literary criticism, short story and essay, textual commentary and image, new and classic texts, share a propinquity, and supplement each other.

This book cannot be allocated to one single discipline. Its theme reaches beyond literature, sociology or philosophy. It is thematically *open* and can be approached in many different ways. It is not tied to topicality, but touches it obliquely — as will be observed in the contributions.

B.V. & M.V.
*Antwerp, September 1993*

# THE AUTHORS

ROGER AVERMAETE (Belgium/1893-1988) — Essayist and cultural publicist of mainly art monographs. Published a.o. *James Ensor* (1949), *Rik Wouters* (1962), *Henry van de Velde* (1963). Wrote dozens of books in both Dutch and French. Became member of the Institut de France in 1981. Wrote two books on Antwerp: *Synthèse d'Anvers* (1932) and *Anvers* (1942)

MARCEL BROODTHAERS (Belgium/1924-76) — Made his début as a poet in 1957 with *Mon livre d'ogre*. In 1964 his exhibition *Moi aussi, je me suis demandé si je ne pouvais pas vendre quelque chose et réussir dans la vie...* did not pass unnoticed. It was the start of a plastic œuvre, language-wayward and very Belgian-iconoclastic, that, together with the works of René Magritte, would dominate nearly all artistic movements of the seventies and eighties, both in and outside Belgium. His controversial *Musée d'Art Moderne. Département des Aigles. Section XIXème siècle.* (1968-9) caused an especial commotion. His compelling attendance at Dokumenta V (1972) canonized Broodthaers as one of the most influential artists of the century.

ANNE CAUQUELIN (France) — Novelist and philosopher. Published a.o. *Potomar* (1978), *Cinévilles* (1979), *Les prisons de César* (1979), *Essai de Philosophie urbaine* (1982), *La mort des philosophes et autres contes* (1992), *L'art Contemporain* (1992).

STEPHEN R.L. CLARK (Great Britain) — Professor of Philosophy at Liverpool University. Published a.o. *Aristotle's Man* (1975), *The Moral Status of Animals*, *The Nature of the Beast*, *God's World and the Great Awakening* (1991).

WITOLD GOMBROWICZ (Maloszyce, Poland-Paris/1904-69) — Novelist and playwright. Published a.o. *Ferdydurke, Pornography, Cosmos, Diaries* (2 vols).

LUUK GRUWEZ (Belgium) — Neo-romantic poet, made his début in 1973 with *Stofzuigergedichten* ('Vacuum cleaner poems'). *Dikke Mensen* ('Fat people', 1990) and the 'siamese' diary he wrote together with Eriek Verpale, *Onder vier ogen* ('Face to Face', 1992), got unanimous praise by critics both in the Netherlands and Flanders.

MARC HOLTHOF (Belgium) — Member of the editorial staff of the periodical *Andere Sinema*, screenwriter of a.o. the television series *Beeldcultuur* ('Visual Culture', Bert Leyzen-prize 1988). Numerous publications on film, visual arts and literature.

KRIS HUMBEECK (Belgium) — Literary theoretician and publicist. Member of the editorial staff of the periodical for the study of Louis Paul Boon *De Kantieke Schoolmeester* and of the literary magazine *Restant*. He wrote a reference work in seven volumes on the steam engine and is (co-)editor of a.o. *De/constructie. Kleine diergaarde voor kinderen van nu* ('De/construction. Small Zoological Garden for Children of Today', 2 vols, 1987), *Discontinuities. Essays on Paul De Man* (1989), *Schrikkelijk spoorwegongeluk* ('Terrifying traincrash', 1991).

MARTHE KILEY-WORTHINGTON (Great Britain) — Biologist. Became what she dreamed to be on the age of six: someone who understands (or at least wants to understand) animals and communicates with them. Did much fieldwork in circuses and zoological gardens. Author and animal consultant.

ULRICH MELLE (Belgium/Germany) — Philosopher and eco-activist. Numerous publications in different magazines on eco-philosophy and phenomenology. Published a.o. *Das Wahrnehmungsproblem und seine Verwandlung in phänomenologischen Einstellung* (1984).

PAUL PELCKMANS (Belgium) — Literary theoretician. His field is the overlapping of literary studies with the history of mentality and antropology of modernity. Published a.o. *Le rêve apprivoisé* (1986), *Concurrences au monde. Propositions pour une poétique du collectionneur moderne* (1991). Editor of *De nieuwe historische roman* ('The new historical novel', 1988).

FERNANDO SAVATER (Spain) — Novelist, playwright and philosopher. Collaborator of *El Pais*. Received the 'Premio Nacional de Literatura'-prize for his essay *La tarea del héroe.* Wrote over forty books, a.o. the recent *Etica como amor proprio, Humanismo impenitente, Etica para Amador.*

PAUL VAN OSTAIJEN (Belgium/1896-1928) — Flemish modernist poet and essayist. Published a.o. *Music Hall, De feesten van Angst en Pijn* ('Feasts of Fear and Pain'), *Bezette Stad* ('Occupied City'). Collected Work: *Poëzie/Proza* (Poetry/Prose, 2 vols).

JACQ VOGELAAR (The Netherlands) — Novelist, poet and essayist. Contributor to the weekly *De Groene Amsterdammer* and the periodical *Raster.* Numerous publications since 1965. Recent work includes: *Het geheim van de bolhoeden* ('The Secret of the Derby Hats', children's book, 1986), *Proces-verbaal van Franz Kafka* ('The booking of Franz Kafka', a reader, 1987), *Verdwijningen. Oefeningen* ('Disappearances. Exercises', short prose, 1988), *De dood als meisje van acht* ('Death as an 8-Year old Girl', novel, 1991), *Sriptease van een ui* ('Striptease of an Onion', essays, 1993).

# The Zoological Gardens

## Roger Avermaete

It pleases man to look upward: he surveys a world beyond his grasp. (Except a few poets from a bygone age, no one has ever dreamt of seizing the stars.) If man chooses to pay attention to his own scientific knowledge, below him there is a world in action whose existence his imperfect senses cannot discern.

All the same, man is king. That goes back to the time he invented the creation. The fable may be puerile, yet he does play the leading part (did he not fashion God in his own image?) and that makes up for a fair number of errors. One does not give up a role one has made to one's own measure. To the moon, the stars! To nothingness, the atoms! Behold man. And the beings that would match him had better beware! He accepts vassals (but only to reduce them to slavery). He exterminates the recalcitrant, except — supreme arrogance — those specimens he keeps in cages.

The eagle, that haughty predator, unable to spread its wings. The lion, that great traveller, obliged to wear out his paws in a spot a penpusher would deem too narrow. And all those others — claws and fangs — known to be ferocious and sanguinary, that you are free to go and scoff at your leisure, protected by solid bars, behind which all those untamed beasts slowly die of boredom, of nostalgia and of impotent rage.

This is just to tell you that, even if you consider it to be in extremely bad

1

*Poster of Antwerp
Zoo (Royal Society
of Zoology,
Antwerp)*

taste, the Zoological Gardens have been invented so that man's omnipotence may be attested by all the species whose representatives languish in its gaols.

Some can now assert that the Zoological Gardens in Antwerp are among the most beautiful in the world. I cannot — thank God! — judge whether this is so. I have never seen any other. My childhood has wobbled there on an elephant' back. I have straddled ponies and dromedaries there. Animals from all the corners of the globe have seen me running and playing in front of their steel bars. The only memory I have kept of them is their display-dummy posing, behind the white enamel label bearing their family name in Latin. And the smell that radiated from them was to me like the very expression of the boredom exhaled by their bodies.

Boredom, sovereign god of the 'Zoo'! From the marabou with its appearance of an old solicitor in its morning coat of faded black, to the elephant which resembles an old gentleman in disheartened trousers. Enormous yawn of the lion finishing in a roar. Owls like well-imitated curios. Giraffes, with a ceiling on their heads, astonished at their disproportionate necks. Immobile tigers you would put just as they are in furrier's window display. A fox — ah, those useless mischievous eyes — in a cubic metre of space. The snake you would mistake for the fire brigade's thick hose. Kangaroos, ostriches, bison. You are perfect models for the animal sculptors. They find you each day in the same place, immobile, stunned.

There are the agitated ones. The polar bears overcome with the heat. The sea-lions that have water to swim in. The monkeys. Great attraction, the monkeys. Is it perhaps natural sympathy? They are so close to us. Their gestures are more impish, but their face is graver. The most sullen of our official expressions appear hilarious next to their mask fashioned for a tragedy whose theme is unknown to us. Are they really cheerful, those agitated animals who amuse themselves without laughing?

It was a monkey who ate my nanny's hat. A beautiful hat, a fair imitation of a vegetable garden.

Only the hippopotami seem happy with their fate. When their large head

*Antwerp Zoo: the human fauna (Royal Society of Zoology, Antwerp)*

4

emerges from the water and they snort complacently they carry the satisfied expression of a fat middle-class couple that can afford holidays at the seaside. In fact, if they were wearing swimming trunks I could give you their name, I mean the name of the satisfied fat bourgeois who resemble them like brothers.

There are not only animals in the 'Zoo', there are also people. I am not talking about the visitors but about the members. For the 'Zoo' is part of an association to which it is proper to belong. Many Antwerp citizens are interested in this branch of the natural sciences. Some do not hesitate to cut down on their daily eating habits so as to have the right to frequent this useful institution assiduously. In fact, the true member never visits the animals. He occupies in that part of the gardens where only the human fauna stroll around in liberty. It is there, around the bandstand, and to the encouraging sounds of the music, that new dresses and original hats are worn for the first time; it is there that the learned assembly of mothers auscultates, judges, decides; it is there that romances and intrigues begin and end. Bait and hooks. Around the bandstand, young men and young girls, turn, always turn — do not fear, sooner or later it will bite! Watch your parents, young men, young girls: they are skilful anglers, at the right moment they will pull up the line and the game will be over. Others will replace you in the circling circle, and you will watch the game until the day when, gently, you will lower a line for your children so that a gentle fish may come and swallow it, a gentle fish of the circling circle of the 'Zoo'.

The visitors are not very interesting. They only have eyes for the beasts which they watch as if they had never seen them before. Which is, after all, quite possible.

What I like most about the Zoological Gardens are the deer and the mouflon: they regularly spit in your face.

*Translated from the French by Ortwin de Graef*

5

# Animal garden for today's children:

# third series

## *Kris Humbeeck*

*An Escape*

Five o'clock. Mr Telleke, though no marquis, leaves the Book. It closes behind him on the endless susurrus of 313 satin ball-gowns on a sultry summer evening. Notary Public Telleke turns the corner. 'I'll find him', the attentive reader still hears this strange person mumble.

This is Telleke, a gentleman, and he's on the move. He's searching. He goes a-searching towards the station — or rather, he is being moved: thirty-three trains inside him are coming and going. Coming-and-going. This gentleman, Telleke, escaped from a Collected Works (Volume III, *Grotesques and Other Prose*),[1] resembles an automaton. People pass him by unnoticed, he pays only mechanical attention to things. Cars glide past with the soft whisper of windscreen wipers, trams screech on their glistening rails, the bells of the old railway cathedral strike 13:02. 13:02? Gentleman Telleke comes to a halt. The trams are dead animals. The cars close their eyes. On the dome of the impressive building a Belgian flag is dripping next to a multicoloured pennant: ANTWERP 93. Only a little girl with a hoop is still out on the streets. All extremely melancholy. 'L'union fait la force', thinks O.D.P. Telleke. And he starts moving again — or rather, something inside him starts moving again. The powerful wheels are thrown back into gear. Trams are sounding their bells, cars

*Antwerp Central Station (Belgian National Railway Company)*

are bellowing, letters are being delivered again. From under the station roof a locomotive screams, O alluring motif. It sounds like an animal from the days of yore.

    Gentleman Telleke enters the station, an enormous arcade. His instinct drives him to the Zoological Gardens. The gardens are next to the station, that's logical. 'In the Zoological Gardens,' hopes something deep inside Telleke, 'I will meet someone. Someone who will rid me of all my anxieties.' That, too, sounds logical, for today's children.

7

Deracination, the staggering multitude of things and an ever-increasing confusion in time and space, we live through it day by day — joining the movement in a valueless world. Ostensibly, we have learnt how to live with it all, in merry acceptance: sure enough an accident more or less doesn't matter! With an ease and a casualness that verge on the obscene, we make use of telephone, fax, train and motor car. Mindlessly, we yield to the civil law, which is the law of progress. And fathers no longer fill us with fear, they are simply there — or quite simply not: who cares?

This is what gentleman Telleke means: delivered from an age-old fear, man allows himself to float on free waves of ether and he tunes in to the vain chatter of opinion-makers as to the truth of the day. One feels secure: 'We are being looked after.' Notary Public Telleke, who is pathologically afraid of going off the rails and regularly falls prey to a sense of complete disorientation, is now looking for his spiritual father. He is a-looking for his father. Oh! oh! that is considerably less logical, in these postmodern times. Yet classical. 'And anything classical is good', says our beloved mayor Fuckle in his office at the same moment. Amedée Fuckle loves what is classical. He is classical, an archetype even. For 313 years to the day he has been the keeper of this city. This high official is just on the point of breaking into a song of self-praise, when his loyal servant Engelbert Druyventruyt rushes in with alarming news: the Notary Public has escaped! Or has he perhaps been kidnapped?

Uniforms rise behind their desks, blue light and sirens take over the port, alien elements are subjected to violent interrogations. Mayor Fuckle is personally in charge of the Operation: 'No matter how classical one may be, the possible abduction of a modern hero is no trifling matter. And van Ostaijen, too, belongs to our cultural heritage, to be sure!' Thirteen out-of-the-blue paparazzi hold a gigantic microphone under the mayor's olfactory organ: 'Presumably we are dealing here with some sort of bad joke, gentlemen. Fi! Those young foreigners had better stay home and do their homework in peace and quiet!' At the other side of the metropolis meanwhile, brown-shirted gangs are running around, plundering, raping and starting fires. Their leader, going by the name of Staf Fillokok, has spoken hard words in the Atupalian parliament. Blood is in the air, the sun sets. 'Van Ostaijen must be avenged!' sounds from 3113 hoarse throats in a neglected suburb. About this Amedée Fuckle is of course not in the know; that, too, is classical. At the moment the first non-Atuplian dies in the run-down working-class district of Orthouberg, gentleman Telleke enters the Zoological Gardens.

*Zoological Gardens and station*

'In the Zoological Gardens and the station', thought Notary Public Telleke, 'the nineteenth century experiences its greatest triumph.' Behind the apery, the stone dome of the railway cathedral rose up high. So much splendour had never been beheld; gentleman Telleke raised his enraptured eyes to heaven and spoke: 'In this colossal house, modern man tames his worst nightmare, the monster incarnate of the iron road. Thus he allays his most primitive fear, the fear of losing himself along the way.'

'CENTRAL STATION', O.D.P. Telleke had read upon entering this sanctuary. The words rose in capitals from his memory, and a beatific smile appeared on the Notary Public's otherwise quite impassive face: everything breathed out liberation. 'The station is alpha and omega of modern man. Here he departs to arrive at a Better World, here he surpasses his physical constraints.' The citizen-schlemiel Telleke, suddenly a philosopher, drew a deep breath and spoke: 'Transcendence, revelation and finality, those are the things man has always longed for. But only with the advent of the train did these essentially human, purely utopian, desires become real. The fairy tale came true, the dream of paradise regained became a concrete project. Like a golden terminus this paradise has begun to shine ever since on the horizon of our history.' Only now did Telleke really warm to his subject: 'The steam horse is history, and our modern history is a secular apocalypticism: man destroys what is, so as to rebuild the world to his own divine likeness. It is called progress. And progress is sacred, our history is a modern eschatology.' Gentleman Telleke approached the apery, the sun sank blood-red behind the station: 'Under this much too high roof we appropriate the world.' Then Notary Public Telleke looked at the ape. The ape, a chimpanzee, looked back in sadness. Intently, Telleke nibbled a nut.

This is the man Telleke and his thoughts start soaring.

In the advent of the train, man lived through his mystical rebirth: he became modern. As a child of God, a reader of the Book of Nature, man belonged as of old to a history which seemed writ-

9

ten in advance, once and for all, only to unravel itself slowly to believers under the watchful eye of a Higher Being. *L'amor che move il sole e l'altre stelle!*

Many years passed by in Notary Public Telleke's mind's eye.

Up until the first half of the preceding century our cities rested in scenic seclusion behind their old, oft-times crumbled, yet still symbolic walls and bulwarks. In the small picturesque port of Antwerp, for instance, which in 1800 counted a mere 56,000 souls, toll collection remained customary well into the century. Antwerp, 1832, the last Dutchman has been chased away. God is in his heaven and peace reigns once more, merry Middle Ages. Not that there were no paupers or destitute around, on the contrary, but riches or needs are not ordained by man — one merely helps where possible. Apart from that, the affluent Sinjoor² drowses and swoons in ancient blandness, revelling in a petty merchant happiness. Steam engines can still be counted on the fingers of one hand. On the ramparts, mills are grinding interminably slowly, to the rhythm of the wind. Within these walls reigns an antique *esprit de clocher*. At ten in the evening, when the citizens pull nightcaps over their ears and only a few dead drunk paupers still hobble down the narrow streets of the city, the entrance gate is closed. All those who still want to get in with horse and cart have to pay a high levy. Sentinel and night watchman keep an eye on things. Military force there is no further need to fear, but fire remains a dreadful adversary. Oh! oh! that red brute. And suddenly the train was here, the fiery salamander: totally and bewilderingly new, producing strange noises, oh! never heard. A green eye — and a red — in the dark. Thunder — smoke — and spark. All very confusing, sir. Please tell me, don't you like to see it lap the Miles? Listen! oh! listen: it shatters an ancestral silence with its shrieks and squeals. Don't you hear the rumbling and rattling in the bowels of this terrible space-gobbling beast? Its hot breath still lingers in the frightened firmament when its thud and thunder are already dying down behind the horizon. Where art thou, Time? Where art thou, Space? It acknowledges no limits, this puffin' Colossus of Steam, pulling into old fortified cities, one by one, in royal triumph, smoking and snorting, conquering the world at a speed which only to us moderns (train travellers in our deepest thoughts) is incongruous with the accompanying stench and dirt and smother. Lo! isn't it truly diabolical, this roaring 'wild thing': it moves too swiftly, accidents will happen ...

Then everything went purple in Notary Public Telleke's imagination. The man of the law pricked up his ears, blew frankincense and myrrh from his nose. Rustling. Crackling. Sizzling. Was there a sound? This is Telleke, a gentleman, he hears Pope Gregory XVI in a piping and creaking voice (as if 3113 knives were being honed at the same time or a train stretching for miles were braking hard) curse the godless steam engine and the iron road: vade retro, Ferrovias! For a moment the beatific smile on Brother Telleke's thought-furrowed face

*Old train (Belgian National Railway Company)*

disappeared; he realized: the creator of the steam horse no longer yields meekly to the quirks of Nature, he violates the divine order of things and spurns heavenly mercy.

What the train-man cultivates, he no longer does in the name of a Supreme Being. In his demonic pride he rewrites the Book of Nature. He makes the most of forces that are only secretly present in the world, that rest inside Mother Earth's belly or are engendered artificially, yes unnaturally. Black gold, the diabolical force of compressed steam, these herald a second Copernican revolution. As a potential ruler of a nature-in-full-movement, man henceforth writes history himself, in his own name. The Central Station, pride of this city, is a visible sign of this.

The ape looked stupidly at Notary Public Telleke. The latter, by no means out of breath yet, resumed his monologue.

The station is a metaphor. It is the monumental idea that man moves freely and on his own, and preferably in concord, toward a hoped-for terminus: his paradise on earth. In the station, where trains ceaselessly come and go, modern man celebrates the imminent end of his own history, his self-made destiny. Here he rises to heaven on the wings of his bourgeois imagination. The station is magnificent. The Zoological Gardens are also magnificent. Station and Zoological Gardens are both magnificent, each in their own ways. In the station, man cancels all distance — the station is order and perfect communication, not counting minor accidents. The Zoological

11

*Antwerp Central Station (Belgian National Railway Company)*

Gardens are also order, but they are of a totally different order. Smaller in a way, more child-like, ridiculous almost. The Zoological Gardens evince a bookkeeper's order, no Egyptian pavilion can alter that. In the Zoological Gardens, modern man is secretly ashamed of the little-ness he unintentionally puts on show and displays as the king of creation — after all, he is in the open air. In the station, a covered space, on the contrary, this insignificant worm imagines him-self a hero, a lion or eagle. It is more magnificent to tame the monster of the iron road than to lock up wild animals great and small in cages or aviaries. Nevertheless, station and Zoological Gardens continue to be inextricably bound up with each other in our imagination as well, like the labyrinth and the Minotaur. The phantom and the opera.

Again the Notary Public looked at the ape. And again the ape looked back, in endless sadness. 'This chimpanzee is sick of the sea', the attentive reader was still able to hear the melancholy gentleman Telleke whisper, before his thoughts once more slipped down to the train.

We're in the latter half of the nineteenth century: Hear! Everything labours, everything sings. A beehive is the Nation — everywhere the soft buzz of industry and prosperity is heard, nowhere do peace and quiet still reign. Like a gigantic web the iron road is spun over the father-land. One big maze of rails, that's what life from now on is. And see how the winged steam steed floats above these glistening rails, hither-and-thither, lo! hither and, lo! thither ...

Here gentleman Telleke indulged in one or two dance steps. It resembled a Viennese waltz and stopped as abruptly as it had started. Telleke, not a little star-tled by himself, plunged into his memory.

Even though the iron horse at first still bent to the oddities and irregularities of the land, now modern rationality wins out. Expropriations occur in massive numbers: houses, indeed chur-ches, are torn down. Soils are levelled and the land is raised. Tunnels are bored and viaducts built. Then railway workers in Fochania dug up a bizarre creature, half ape (an ape in Fochania?), half human.

Petrified, gentleman Telleke, a one-time dancer, stood before the pit. In the pit was something that should always have remained hidden, a grotesque mon-strum, barely five feet, with enormous cheekbones and an excessively flat head. 'How incongruous,' sighed the old man Telleke, 'a misprint in the Book of Nature!' He closed his eyes in terror. Lime and loam were supplied, the mon-ster was hidden from view. To this very day the rails on this spot swerve strangely.

But what does a small layer of earth avail: man is curious, his memory fatal. He is unable to efface the recollection of the ape-man. At night he wakes up, bathed in sweat: is he an animal, then! Never again will man know peace if he cannot find a solution to this problem: for his divine origin, and with it his whole being, is at stake.

Man racks his brains — not since the Sphinx had his thoughts been so profound. He looks, and looks, and finds a solution. He demotes himself to ape — oh! to a worm or a fish if need be — only to be able to ascend heaven as an *Übermensch*.

This quasi-oedipal ambiguity he acclaims with all his dialectical ingenuity as evolutionary principle: this is where I come from ('from still black waters deep under the earth'), that is where I am going ('like an eagle straight upwards'). With this disposition man also studies his body and woman's — which he experiences, in all its bestiality, both despised and fondled, as an essential shortcoming — only so as to ingeniously escape this corporality, the shortcoming incarnated by woman, and shame: he is no beast. Or is he? Man hides his shame, literally — especially that of woman, eternal keeper, like it or not, of an ineffable secret. Much is suppressed and forgotten by man, much he declares sacred or taboo: for not everything is meant for the ordinary ear or eye. But look, these peculiar blades on our backs: used we not to have wings there? Oh! oh! man wants to become a bird of paradise, the mythical creature he used to be before his imagined 'Fall' to earth. Seven walls were pushed away, much became suddenly clear to the archaeologist Telleke:

That's why modern man scratches about in the earth so much, to be able to flee it as quickly as possible. That's also why he invents sources of energy that have no place in the Bible and why he resolves forces whose origin remains unclear, whose effects are incalculable — so as to transcend this world as fast as possible, and with it his own, imaginary shortcomings. Chimpanzees are sent into outer space by modern man — then he himself leaves. Feverishly he feels his way, away from this terrible labyrinth. He sighs but for one thing: the world has to become orderly and transparent again, a second Eden.

Then, Notary Public Telleke dwelt in glass buildings entrussed with cast steel from which was suspended, shockproof: future Man, a computer-controlled winged phallus. 'We must abolish lack', mumbled Telleke.

To that end knowledge is required: where do I come from, where do I have to go to? But along with this knowledge, confusion and insecurity grow, too. Everything turns to ashes and is blown away. In his laboratories, nervous man searches for the ultimate antidote, a remedy against, excuse me, for life — and its endless diffusion: the canker of our daily existence. But modern alchemy disappoints and in the heart doubt rises. Man, therefore, to set his mind at ease, builds Zoological Gardens. He means: look, the animals have already been conquered — we're on the right track.

Gentleman Telleke fed a nut to the ape, which was in violation of all the rules of the gardens. Not a mouse stirred, only the zebras got nervous — or was that an illusion? O.D.P. Telleke, innocently, wound up his argument.

In the Zoological Gardens man reigns like a demigod over the world, his private domain. He manages his own history and structures the endless differences to which this history gives sense, aim and direction, with an ease and an obviousness as if only one combination of the so-called facts were thinkable, only one narrative. Man says: Behold, that is an ape — and this is me. From that ape I descend. At the same time, in Zoological Gardens this modern man draws an advance on his promised paradise. He confirms himself as the Other, who is yet to come: the mythical god-man he would have been once. He sees a bird majestically flying heavenwards, higher and higher, apparently weightless, moving in the welkin. Man realizes: this is the way to do it, freely and on one's own, totally of one's own accord.

Lo! here is a modern man, Telleke, he arranges things in accordance with his own views. He imprisons the wild animals and studies them — all quite objectively. If necessary, he cuts the animals open: he will fathom! Then he sets the seal on his work and builds, next to the Zoological Gardens, a station. In this station the beast of the future lives, the vapour steed. Like a god, man has created this animal out of iron and water, to conquer space. The train is the likeness of man. It is reliability itself, even if in railway accidents everything returns, that modern man in his youthful recklessness represses and denies. Yet, in spite of all minor accidents and derailments, this man-beast is not bound by nature. It can handle all climates. It links coasts and continents, bridges mountain passes and explores the darkest heart of Africa. When it sleeps, it is because its master is tired. It is itself indefatigable.

Then night fell. In the station dome, the last lights were extinguished, the Zoological Gardens were sunk in a sound sleep. Even the night watchman was snoozing. Not at all surprised, Notary Public Telleke found the door of the apery unlocked. With an infectious yawn he laid his head down to rest. 'It all began here', the attentive reader was still able to hear this strange person mumble, before he entered another world.

### The first day: yet another escape
That first morning something strange happened. Somebody must have been scandalously negligent, for a zebra had escaped its cage and in the morning rush had run onto the street. Across the entrance of the Central Station the animal lay oddly folded up underneath a small, bright red Japanese car. It seemed asleep, its head peaceful on its broken forelegs — not a drop of blood in sight.

The lady owner of the conspicuous vehicle stood watching and crying. Typists buried their faces in their hands, mothers roughly pulled their pointing offspring from the scene of the calamity. The tormented policeman Benny Dorm, who had seen it all happen, wrote out a ticket in his logbook, his seventh already this week. The Zoological Gardens opened their doors half an hour later than usual. Apart from that it was a summer's day like any other. Notary Public Telleke, woken up by the dogged sound of an ambulance, had, upon seeing the first keepers, retreated into the apery's deepest darkness. The somewhat full-bosomed lady who had been feeding the animals had failed to notice him. These were holidays. Trains were ceaselessly supplying tourists. Colourful balloons and children's cries filled the gardens. Gentleman Telleke peeled a banana and, as always, lost himself in speculation.

'Well then, here I am,' Telleke thought, 'among primates. And nobody noticing anything, because that's how I want it.'

From under the station dome a metallic voice announced the departure of a train on platform 13. Lost in thought, the Notary Public moved along with the many-headed multitude. Barely had the platform started moving when an indescribable panic seized the man of the law. The train was picking up speed, Telleke went wild. The rattle, hiss and groans swelled in his ears until they sounded like an apocalyptic cacophony. 'I'm jumping', Telleke thought. Just as he was about to yank the window open, he was woken up by the loud giggle of a giraffe, the funniest animal in the entire zoo. Gentleman Telleke opened his eyes and stared straight into a schoolboy's astonished face (freckles, squint-eyed, a check cap on inevitable straw hair). He was no longer standing in the dark. Here is a man, Telleke, he is standing in an apery and sticks three fingers between the bars. The schoolboy offered him a banana. Telleke, nonplussed, took hold of the banana and wanted to say thanks. He changed his mind, unconsciously pulled a jolly face and peeled the fruit in silence. The bystanders laughed, relieved it would appear — or did gentleman Telleke imagine this? 'Well then, here I am', thought Telleke for the second time, 'among primates. And nobody noticing anything. Because that's how I want it?'

*Antwerp Zoo (Royal Society of Zoology, Antwerp)*

## The second day: station and Zoological Gardens

A shock ran through the Cultural Capital of Europe. The day before, the bizarre accident in front of Zoological Gardens and station had been amply commented on in its many taverns. And that morning the unfortunate zebra appeared to have made the front page of both the somewhat popular *Antwerp Daily News* and the more serious *Atupalian*. Many happened to be reading the journalistic explanation of the facts when the national radio service reported a second accident. Same protagonists, same scenario: a perfect repetition of history. All very uncanny. The shudder running through the city was surpassed only by the tremor produced in the local telephone network by Mayor Fuckle's powerful timbre. Antwerp's high repute was in danger, 'Right now, when all the eyes of Europe are directed at us. As if that van Oskade case weren't enough!' A terribly tense managing director of the Zoological Gardens once more summoned his personnel: the culprit would be punished severely and in ways that would set an example. But who was the culprit?

Totally innocent, in any case, was Notary Public Telleke, who had not even been woken by the tumult in the street. He woke up when a monkey nut hit his brow none too softly. He opened his eyes and descried a small commotion around the zebra cage: it was long after noon, he had slept far into the day. Here was a man, Telleke, lying in the dark of an apery. He had dreamt how he was standing in the full sun, three fingers between the bars that separated him from the others, gaped at by an astonished crowd. Then, a schoolboy had offered him a banana. He had wanted to refuse but been unable to do so. A strange dream. Or had it all really happened like that? Notary Public Telleke could not recapitulate the past day's events. A ray of light fell through the rails in the cage, the rose windows of the railway cathedral reflected the sun. Zoological gardens and station had not only been built on each other, they kept in touch in more mysterious ways as well.

'These Zoological Gardens, this station, they differ from everything man had been dreaming of before', judged gentleman Telleke (though a level-headed man).

To be sure, before the advent of the train, too, man took pleasure in ambling around animal parks. Oh! all along we seem to have given in to the urge to prove our strength over the other animals by locking them up and displaying them. From time immemorial their sumptuous captivity is taken to be a sign of our divine descent and historical destination: Thou wilt subject nature to thyself! More than 3000 years ago, the Chinese Empress T'ang Ki had instructed her subjects to build a marble Home for her Deer, and the no less powerful Wen Wang, some twenty centuries later, had the vast Ling-Yu park laid out: the Garden of Genius. Genius in those days still used to mean superhuman wisdom, which the famous Solomon, too, possessed. Solomon was a zoologist, as was Nebuchadrezzar II, who was also called Nabu-kudurri-usur II and not only had the Hanging Gardens in heavenly Babylon but also an entire menagerie. On the subject of these wild animals and the like, Aristotle wrote: Here all being is revealed. Or was it Heidegger? In any case, the Stagirite's most combative pupil was Alexander the Great. Many a tribe there was which he crushed, looking for acquisitions to fill his world-famous animal park. Thus, from the beginning our rich tradition linked Zoological Gardens, divine truth and power.

Once again, many years passed by in Notary Public Telleke's mind's eye. He visited the royal zoo in Rome (a she-wolf's gift) and watched the freshly crowned Charlemagne in Aix-la-Chapelle put a mouse on trial (he would accept no denial). 'Philip VI bred foreign rabbits in the Louvre — he went bananas — but the oh so well-balanced Bourbons introduced piranhas to the ponds of Versailles!' In Schönbrunn, where later Romi Schneider would live, the royal throne was said to have been defaced by the droppings of exotic birds. (Oh, that's very nice, isn't it!) And on the red carpet in front of the palace an iguana is reported to have drowsed, while alligators were seen in the large drawing-room eating lackeys. 'The Vatican, verily, it is too little known, was a second Noah's Ark for centuries on end. Animals lived there, ah, from nearly every country.' Woe to us! Have we understood the man of the law correctly or not: was there something going on between Gregory XVI, the female babiroussa and a dwarf elephant?[3]

Then the train came.

It is 27 September 1825. In ways never before beheld the Locomotion bridges the distance between Stockton and Darlington. And look: the whole world starts moving, everything revolves and turns with a will. Was this the beginning of a New World? Telleke, a professor now, lectured:

Certainly, even before the coming of the steam horse, reason had already been called sovereign and the world was a wonderfully glistening machine. But a machine, so to speak, which did not really move, nor set anybody moving: cogwheels set with superhuman precision, a clock. The maker of that clock was God, in this way Newtonian man settled amicably with his heavenly father.

We're in the year 1700. At midnight, when the villagers are asleep and only a few lovers are still rustling in the thicket, the researcher sits in his garden. Through his telescope, an articifial eye, he watches things coldly; nothing distracts him. Not the lovers' rustle, nor the hoohoo of the Carine Noctua. In the firmament, 313 bright stars stand stiffly in line. Like splashes of paint frozen in space, poetized the morbid poetaster, oh! 313 traces of a god touching himself a little too lewdly. Yet, what would poets know these days? The new man has learnt to see the world differently with the aid of his lenses and microscopes. More sharply, more precisely: flowers, books, beasts, people, planets, bells — the whole of 'nature', it is no impenetrable cipher any more than it is an insane scribble. God is no dauber — nor is he a lecherous bear — he is an abstract genius. Descartes already knew, the letters from the Book of Nature are numbers and these numbers, to those able to read, form significant combinations. So it is with the stars, too. Look, there the Pole Star stands, bright and clear-cut — that's the anchorage. Once it has been discovered, the rest follows on fast: a whole zoo! O yes, these heavenly bodies move, like our very own Mother Earth, yet they do so in fixed orbits. Of course, there is variation, there are differences — in time and in space. But those differences constitute phases or periods, thereby allowing space to be mastered in time. All is eternal recurrence and identical repetition, hence perfectly calculable: from the apple falling from the tree up to interplanetary dynamics. Devious leaps, quirks and whims are not what things indulge in; they never swerve for no good reason from the course which the Lord prescribed for them at the time of Creation. There is a single iron law, a single ingenious programme. And every exception to the rule is a clue from above. Yes, yes, the cosmos is a clever mechanism, the great heretics already knew that. Copernicus, Kepler, Galileo and even Newton, they all sang in unison: the Great Clockmaker may well have retired, but he has done an expert job!

Then the earth trembled and doubt crept up on the enlightened man Telleke:

For if the world is a big clock, and God the intelligent maker of that clock, perhaps then, God be praised, man is the only creature in the whole wide world that knows this clock to be a clock and is at all able to learn how to read this clock — but as a reader of this clock, clever man must unfortunately always stand outside of 'nature' ... as opposed to the delightfully stupid animals (soulless mechanisms). But then, for the first time in history, thinking man and the animal are creatures of a totally different kind, bien étonnés de se trouver ensemble on this planet — thus science and knowledge sever a time-honoured, 'natural' alliance between all creatures great and small: the illusion of an organic community founded in God. Instead, we get an equally illusory model of cosmic harmony, the delusion of a divine architect or clockmaker.

Here gentleman Telleke paused for a moment, as if he wanted to let the consequences of this and that sink as deeply as possible into his Notary Public's soul.

Then he spoke:

Can it come as a surprise, therefore, that in the blinding light of these classical mathematics, physics and astronomy, the Zoological Gardens shrivelled into a curiosity, a playground for monarchs? And can it come as a surprise, furthermore, that the common man remained insensitive to the cold splendour and iron systematics of things, of which, as a working and stupid person, he was, after all, no part, nor could ever be? Can it come as a surprise, finally, that this gruff science did not really leave the study and never really changed the world? Indeed, does anyone truly wonder why this 'inhuman' theory of cosmic abstinence was not pushed to its furthest limits even by its most eminent representatives and the most brilliant mathematicians? Descartes, Galileo, Newton: which one of them did not honour the Creator, in the *Vanessa cardui*, chancing to flutter past as a tender father, the author of the ancient Book of Nature, rather than as the retired inventor of a complex watch?

We're in the year 1700, it is 7 o'clock exactly, the sun rises — excuse me, again the earth has made half a turn around its imaginary axis. Still sitting in his garden, the researcher. (The lovers and all Strigiformes are asleep.) Patiently, this researcher notes down the coordinates, then adds everything. To be sure, the real fact of the matter and deeper meaning of things remain hidden from the exact scientist — as a physicist or an arithmetician he remains outside these things — but apart from that everything turns and revolves as it mathematically should, in this best of possible worlds. Constancy, regularity: these are enough for the number-loving man. It is entirely reliable, the-clock-the-world. Lo! never before had man feigned so much metaphysical peace of mind. On the ramparts of the old fortified cities, the mills were grinding interminably slowly, in accordance with the eternal laws of classical dynamics.

Then the train came.

And there was movement, real movement, as well as hurry, oh! chaos even: life became one big mess with all this vapoury violence; it almost looked like the Day of Judgment. Panic broke out and the panic was great, for lo and behold: the monster of the iron road had not even yet torn man loose from his heavenly father when abysses opened under his feet and never-suspected evils spread. Gentleman Telleke, bell-wise and loud-voiced: 'Hear, O hear: the hellish fast-coach races like mad over the shins, insanity lurks in every corner. A primeval law has been broken!' Under a pale moon the dead raised themselves from their desecrated graves with indignation. They were not given the opportunity of wailing much: in front of their baffled eyes an old man — his name was God, he had a long beard — ran to his death under the onrushing Colossus of Steam. Lamentations ascended into heaven. But behold: the train became God, and with that train also man. O.D.P. Telleke, visionary, spoke:

Ever since that remarkable transubstantiation, train and man have been steaming together to the terminus of their dreams: a world without unbridgeable distance or immeasurable difference. A homely world, that is the goal of our history, a world without unlit rooms. And steam means light, is source of victory: triumph of thermodynamics! Thus everything is on the move, oh! everything is moving — even the species, which shortly before, Linnaeus and Buffon had still believed to be capable of fixing forever in Latin. Man, behind his compartment window, starts having an eye for alteration, change, variation. Apparent transitions and transformations of the species mix up all existing classifications. In short, the whole world is upside down, the watch-maker's industry is undergoing a crisis.

This is what gentleman Telleke meant: only with the monster of the iron road did one get serious about sovereign human reason, which philosophers and scientists had already devised in their studies centuries ago. The train came and the world unfolded in an endless series of differences, which in addition turned out to be mobile. A new organizing principle imposed itself: there was no longer room for a divine revelation or any other form of predestination and higher guidance. Therefore, man (an engine-driver in his deepest thoughts) took God's place. For the first time in 'his' history would he be creating history himself, in his own name. Freely and totally of his own accord, with open eyes for things, he would be giving meaning to the differences which were being revealed every day between things and their immanent dynamics or logic. In this way, the difference between man and ape was no God-given fact any longer — it became a challenge: Thou wilt free thyself of thy beastliness, Telleke, that is thy mission. Turn thyself into Man! To that purpose, of course, knowledge is required: Where do I come from, where should I go? The old world, conse-quently, shivered with curiosity and a civilizing fever. Zoology turned into a trail-blazing science, the need for space to observe and experiment made itself felt more strongly every day. In the latter half of the previous century alone, over forty Zoological Gardens were opened in Europe, first in London, three years after the triumphant ride of sorcerer's apprentice Stephenson. More than

22

ever the Zoological Gardens became a symbol of power and truth, albeit a truth which civic man wanted to derive solely from his own ingenuity, and analyse singlehandedly, i.e. preferably with scalpel in hand.

O.D.P. Telleke, a chronicler of great renown, got up to speak again:

It is the 5th of May 1835: pulled by three locomotives, including the Elephant, the first train on the European continent steams between the capital Brussels and the archiepiscopal See of Mechelen. In the country's largest port a station is built beside the ancient walls — all very symbolic. On the 3rd of May 1836, with the usual added lustre, the section of track between Mechelen and Antwerp is inaugurated. Three festive trains take King Pamelum I, imported from Teutonia, and 312 other dignitaries to the proud city on the Scheldt: for no less than three full days would the festivities last, after which time no woman remained a virgin.

Thus, enthusiastic Atupal began the conquest of space. Barely seven years later the Zoological Gardens were founded by the Royal Society for Zoology. Meanwhile the walls and ramparts had already been torn down in various places. Meir[4] and station were connected by means of a large road. The city woke up with a jolt from an age-old doze. 'Hear! another breach is being made, they are making a gate and putting tracks through.' Thus, on 15 October 1843, the famous Iron Rhine was put into use: between Atupal's most important port and Cologne, trains were coming and going now all the time. Ever more vehement also were the movements of the steadily growing masses, the world was in full expansion — not a young woman in the whole of Antwerp that wasn't pregnant. Houses and sluices were built, quays constructed. In Telleke's imagination of things, sailing ships were transmogrified into steamers — Antwerp broke from its sixteenth-century straitjacket and in a few years regained the reputation of a seething metropolis. A Golden Age appeared to have begun, more magnificent even than that of Rubens. Oh! ever higher the Sinjoor was pushed up in the movement of peoples. A pinnacle was reached in the building of the Central Station, Antwerp's third railway station, which had required, from 1898 to 1905 (so, for seven years), the unremitting labour of 3112 men and 1 woman. Equally around the turn of the century, the 'zoo' was greatly expanded, at least seven times, it is said.

Thus, station and Zoological Gardens grow side by side, as they well should. They confirm a promise already made to humanity on 2 May 1885: Antwerp would become a truly international port, oh! a global centre of industry, commerce and culture. Millions came to see it for themselves, when the first of three world exhibitions in Antwerp was opened. The Room of Engines was praised by the whole of Europe and admired agape. Yet it was not the steam engine that was the greatest miracle at the time, but the Congolese Pavilion: for there, taking up residence, was a king — Massala was his name — with his suite and seven tigers. He loved us, this pleasant savage. He brought us a little gift, a gift for our zoo. Oh! who in Atupal had ever seen such a thing: a vertically striped horse, with bushy tail plume to boot!'

## The third day

How it could have happened nobody was able to tell, but the facts were undeniable: a slightly battered small red car, a moribund zebra, a totally overcome driver. 'The third okapi already,' cried Amedée Fuckle in his office, 'they're dying out, for Christ's sake!' Electric communication was not required this time, the yell hung like a thundercloud over the city and ere long Druyventruyt could report to the disgruntled mayor that the director of the Zoological Gardens had handed in his resignation. He got his discharge, honourably even, and an official investigation was set up. 'Criminal intent is involved in this,' said Fuckle, 'a thing like this even our age-old opposition wouldn't be able to come up with. The danger comes from outside, although it's inside: fifth column, I know the stuff, I've been through the war. And where the eff is Telleke?' That day the Zoological Gardens remained closed to the general public. When gentleman Telleke woke up, his eye had to forgo the gardens' invariably interesting dynamics. Some four guards in blue uniforms kept watch, one at every corner of the park. Notary Public Telleke scratched his armpits. 'I'm monkeyfying: I've been in this cage for too long', thought the civic man of the law, who by now had come out of the dark completely.

There was no one who saw him; because they did not want to?

The death of the third zebra had seriously upset Atupalian spirits. Only the seven-part TV series 'The Righteous Judges', many years back, had been able to disturb the people to the same extent. The suspicion of a terrible secret frightened the citizens. Fear and a well-nigh pathological excitement were locked in battle within their psyches. The aforementioned Staf Fillokok, leader of the Atupalian Block, took clever advantage of this collective unrest. With a master's hand he surpassed the current incongruities by way of a number of new absurdities. He puffed out his chest, cleared his throat and addressed the parliament in this fashion: 'The murder of our zebras is an attack on everything we hold essential: the enemy is already among us. Let there be defence!' While uttering these final words, Mr Fillokok jocosely swung a baseball bat. Though loath to admit it, the rival factions in the Atupalian Lower House were not a little impressed by this short but forceful plea: after all, something had to be done, the Nation was in danger. For was it not necessary to resist as one man those increasingly insolent foreigners, whose younger scum so plainly did not give a damn about its school work? In the name of political decorum, the widely respected Doctor Wybau ascended the pulpit to demand in a loud voice 'that the laws on school work be strictly obeyed in the future.' 'By everyone, of course,' the ingenious man added, 'without exception!' And he concluded: 'Measures to this purpose should promptly be taken, in view of ... for the sake of ... etcetera.' Moreover, it was 'in this matter scarcely apposite to leave the initiative to Fillokok and his likes, who would actually prefer to have the non-Atupalians not use our own school desks at all, by Jove!' But what did the average citizen actually say? What did his representatives believe he believed? In all party councils this question was earnestly discussed.

Given all the differences of opinion about what course to take, it was eventually decided to hold a referendum that very same day. Post-haste, polling booths were dragged in, two of them still containing skeletons from a previous electoral battle. The turnout was poor. Around midnight it became clear, 33 per cent of the votes had gone to the Atupalian Block: this was a Signal. The morning after the polls a great despondency had come over many people. Even greater was the row that broke out in the afternoon among the politicians. The underworld, on the contrary, remained silent, the Dow Jones did not budge. By the evening parliamentarian spirits had already calmed down: a Royal Commissioner was appointed. Following this, the members of the cabinet and their paladins got into their warm beds: the day had been long and heavy. In the apery gentleman Telleke heard the last train pulling in. It sounded like the groaning of a mortally wounded animal.

## The fourth day

Something of a new habit started growing up in Atupalian life — one got up, washed and dressed and, giddy-up, three flexions of the legs, twice round the table and then the radio on: 'This morning in Antwerp, a zebra ran under a car. The animal had just mysteriously escaped from the zoo.' One drank one's coffee, opened the newspaper and then folded it up again. 'Prominent politician mixed up in financial scandal.' One put one's shoes and coat on. 'Atupalian Block denounces corruption around Zoological Gardens. President of Farmers' Union accused of hormone swindle as well as copro- and peridromophilia. Crooner from Leuven run over by High-Speed Train. Soccer hooligans found new party.' One wished one's family a nice day. 'Taxes and levies drastically increased. The weather for today: fair.' One pulled the door shut and started for the tramstop, whistling. 'The Meuse at Liège: severely polluted, seven stop-logs raised.'

In the Cultural Capital of Europe meanwhile, problems were being tackled scientifically. Thirteen bespectacled experts displayed touching unanimity in explaining the deviant behaviour shown of late by the local zebras: no doubt the animals were propelled by the same suicidal impulse that is known to get hold of whales occasionally, washing them ashore on our beaches in such massive numbers. This superbly unsatisfactory explanation lured many into the Zoological Gardens, where it became more crowded by the hour. Entire hordes besieged the gardens. Two things struck Notary Public Telleke: he himself was increasingly being looked upon as a curiosity. And the steadily increasing visitors acquired more and more animal traits. Or was the latter sheer imagination? Here was a man, Telleke, sitting in an apery and displayed to the people. Laughing and shouting children pointed at him, fathers patiently offered their wives and offspring explanations, photographic labs would prosper shortly. On the dome of the railway cathedral fluttered a Teutonic lion and a Fochian cock lustily in the wind, next to a rain-bleached pennant, the letters of which had long become illegible.

## The fifth day

The day looked like any other one: that morning a zebra had escaped and, almost traditionally, run under a red car in front of the Central Station. Except for the driver of the car and officer B. Dorm, hardly anyone seemed impressed anymore. Arm-of-the-law Dorm, who had seen it all happen again as if by appointment, took things personally this time. Dorm investigated, because 'one doesn't miss a rendezvous with History, after all.' It was he, too, who made the gruesome discovery: in the snake pit, our well-loved Mayor Fuckle — stone-dead, brutally murdered, couscous in his ears and on his bald pate a crescent in fluorescent red paint. Lo! that was not classical at all. Atupal was in commotion, the Palace ordered an investigation.

Unfortunately, however, no indications were found, which proved: this might have been the work of enthusiasts, but certainly not amateurs. Granted, it had been necessary to carry off an over-zealous inspector with a broken leg, after an awkward fall over a blood-stained baseball bat. Telleke, listless against the bars, ate a banana and looked upon the farcical show with sorrow. The whole business looked like apery to him. 'Monkeying around' was the phrase that in turn suddenly occurred to the valiant officer. Benny Dorm had a Span-ish name, but his father was Teutonic. And as dead as a dodo, for seven weeks now: a stroke. His Maltese mother — Veronica was her name, they called her Vie — he had never known. Do these circumstances perhaps explain Dorm's susceptibility to mysterious connections, or was this simple traffic policeman no more than a frustrated sleuth? Did he read too much Eco or Peirce at an impres-sionable age? We do not know, it has never been investigated. Anyway, Dorm followed a trail. 'There's more to these events', he felt, 'than meets the camera's eye.'

That same evening the Atupalian news announced that Staf Fillokok had set up a select emergency committee. The saviour of the idea of the state also appeared live on screen. 'We no longer want to be made a monkey of,' said Mr Fillokok to his viewers in the name of these same viewers, 'a New Age has begun.' He planted his muscular arms on his hips and pushed his chest at least half a yard forward. The speaker's eyebrows formed a single line, and from beneath this bushy frown he distrustfully stared into the living-rooms. The oth-erwise oh-so-gossipy community kept mum: nobody protested, everything was swallowed — for once taxes were not raised. The Leader sensed the whole damned apish country was on his side. Then the calamitous Mr Fillokok made room for the weather forecaster. His name was Bob. He predicted 'Fair weather for tomorrow.' That, too, was swallowed. The next morning, it was raining; by noon the country went bankrupt. It struck gentleman Telleke that not only outside was there a larger crowd than ever, the cages also seemed fuller. The Notary Public himself, actually, had new company. It was the loyal servant Druyventruyt, who sat in a corner quietly grieving. Was it Druyventruyt's endless blubbering that put people off? In any case, a much better draw turned out to be the zebra cage, one day ago still almost empty, now peopled by some five alien elements. Together with the two remaining zebras they attracted a lot of attention, as sole remaining Atupalian specimens of this exotic variant. In the early morning their compatriots had been shipped by the new government to their countries of origin, it was announced. Even if the country had gone bankrupt, the State flourished. At long last it had been delivered from these foreigners and everyone went nicely back to doing his homework. To celebrate the new regime, the Zoological Gardens that day remained open for many more hours. Deep into the night trains came and went.

### The sixth day: animality is in the eye of the beholder

To put it simply: the situation in the Zoological Gardens was swinishly bad. The entire country, moreover, was at its last gasp. But appearances were deceptive.

Morning ritual: a truck carried away the sixth zebra's cadaver, cars resumed their rights, contentedly grunting. On the pavements reigned a holiday mood. The air smelt fragrantly of exhaust fumes, sun-tan lotion and warm blanket. Unrelenting was golden Phoebus, the tiles were boiling hot; small, uniformly blue airplanes pulled the Leader's Proverbs and advertisements for soft drinks across the welkin. In short, God was in his heaven and everything seemed spiffing. In the mayor's sunlit office, too, a great serenity reigned; specks of sun were doing ballet dances in front of the open window. In the corner of the subdued room stood a baseball bat and a hockey stick, on the wall hung a still life with bananas. The chair of the city's first citizen had lately been taken by Gustav Bonobo, the sportiest member of the Atupalian Block, also known as fat Gus. Not because he had such a robust build, but because he never ceased to lay claim, *ad nauseam* even in the opinion of his best friends, to more *lebensraum*.

*Antwerp Zoo (Royal Society of Zoology, Antwerp)*

29

Gustav Bonobo had received training as a tamer of wild animals, and there was no denying this: wherever he passed by nobody dared breathe a word. He was listened to in all meekness and on leaving everyone courteously doffed his hat, even those who did not wear one — more than ever the gesture mattered. The motto of this person in charge? 'Words only increase uncertainty.' Mr Bonobo was a man of action. He seldom talked and only in slogans. He had barely taken the baseball bat in his hands to crack a tough nut on his oak desk, when his slavish servant Hannemans brought alarming news: convincing evidence had been found against the over-zealous officer Dorm, he must have killed the widely mourned Amedée Fuckle!

Notary Public Telleke rubbed the sleep out of his eyes. He despaired of ever really waking up again. Exotic birds warbled madness in his head, a cloud of tiny black flies blocked his view. When at long last sleep had been rubbed away and communication restored, the cages in the Zoological Gardens appeared to be chock-full. Some were literally bulging: arms and legs, as well as an occasional crutch or walking-stick, were sticking out between the bars. The staff meanwhile were in full swing. New accommodation was being hurriedly constructed. In the jostling and mad bustling of mothers and children, uniformed men were bringing up heaps of stones and mortar. The Notary Public was unable to find a pattern in this: it was all turbulence. Only by noon did the dynamics seem to become somewhat more orderly, though the number of visitors kept increasing — or did gentleman Telleke imagine so? In particular the one-time apery was a remarkable success that afternoon: it was the axis of an, as yet, inarticulate revolution. Notary Public Telleke carefully observed all movements in the gardens. It seemed to him there was a mighty concourse, not only outside but also inside the cages. For instance, Druyventruyt had already disappeared — in his place was a B. Dorm, policeman. The two men of the law had barely exchanged a word. Dorm was asleep, in foetal position, his head on his logbook. Contrary to expectations, however, Telleke was more awake than ever. He observed how, with combined forces, a hole was being made in the wall of the station's side. The rails of the already demolished platform 7 were being extended to the middle of the park, where until yesterday the Egyptian Pavilion and big apple tree had shone. Instead, there now stood a strange concrete construction, a small windowless shed or barn that supported a huge chimney, at least seven yards high. What happened inside that mysterious house nobody knew, not even the most heavily armed guard. Then, Telleke had a vision.

Lo! here is a man, Telleke, he is dreaming an evil dream. He is running straight across an immeasurably vast shunting-yard. In his left hand he clasps a railway ticket, his right hand he keeps stretched out in front of him, the fingers spread, as if he were blind or stumbling through the dark. In this way, gentleman Telleke is moving between trains stretching for miles. Invisible creatures are hammering against the doors of rusty goods carriages, barbed wire cuts the Notary Public's face. Then a shot rings out. The sound is stretched endlessly, as at the end of a movie or stage-scene.

The gate swung to with a deafening bang, the park was closed an hour early: there was a distinguished visitor. Surrounded by seven gorillas, Staf Fillokok entered the gardens. He was flanked by Mayor Bonobo, the refined fringe-bearded ephebe Hannemans following in their wake. Silence all around, hanging like a crystal apple in the Zoological Gardens. With a slightly springy step, almost swaying his hips, the Leader proceeded to the apery, straight to gentleman Telleke. Here is a frightened man, Telleke, he has been sold: he awaits his sentence, likely to be carried away in a windowless train to a windowless house that supports an oversized chimney. From that chimney seven inconsiderable clouds will escape. The Great Helmsman (extremely hairy, flat nose with large holes and preliminary signs of ear scabies) looked deep into the Notary Public's eyes. From a minuscule box he spurted a hefty jet of refreshment into his wide-open mouth, drew a few cubic yards of air into his lungs and seemed intent on pounding his chest manfully. But he thought better of it. Instead he pursed his lips coquettishly and spoke: 'See, human being Telleke: this is me — and that is you. We're on the right track.'

### The seventh day

Gentleman Telleke wakes up. The orang-utan lying next to him has fraternally put a mighty arm (paw?) round his hip. At the other side of the gardens, the night watchman mounts his bicycle, while the morning shift arrives cheerfully whistling. On the dome of the railway cathedral the Belgian tricolour flies again. Only the flag of Cultural Capital has been lowered: the world seems back to normal. Had everything been a mere chimera? Notary Public Telleke decides to leave the cage before even the earliest visitors arrive. The apery door is ajar, like that first evening. All animals in the gardens continue to sleep, only a lonely zebra wakes up for a moment. Unseen, gentleman Telleke escapes from the Zoological Gardens and turns the corner. The first trains arrive, the station spits out people. When Mr Telleke steps on the zebra crossing, tyres screech — accidents happen so easily! A van with black-and-white stripes had shot the lights, a lady was the victim. People come flocking round. An officer — his name is Benny Dorm — tries to keep the chaos under control. Gentleman Telleke does not see it, he is being moved by a mysterious force. He enters the station. He takes the underground. Lo! he boards the tram to Schoonselhof. The tram is a nineteenth-century model and packed.

Here is a man, Telleke, he is standing on the platform of a packed tram. Under him the glistening rails are gliding by. An entire world is moving. That world is searching. It moves a-searching to what must come. There, in the early-morning mist, the municipal cemetery looms up already, the end of this history is dawning. In this garden, like a sacred vow and held in marble, rests the lifeless body of Paul van Ostaijen. But how to describe Notary Public Telleke's and our own surprise: the stone in front of the grave has been rolled aside! Something steps forward out of the dark. Here is a man, Telleke, he stoops forward and sinks in sacred fear on his knees. Seven zebras come from the Prince's Grave,[5] it's a miracle. Their eyes are like lightning, the white of their fur is whiter than snow. O.D.P. Telleke quakes with horror before them. Afraid, he huddles up against the great writer's final resting-place, shivering he lends an ear. Gentleman Telleke turns into a statue, stately and classical. Then black angels break into a requiem. The supreme angel speaks:

Fear not, Telleke. For I know thou lookest for thy spiritual father, the first writer of this town. He is not here, for he has been betrayed. Thou also shalt betray him while looking. Therefore, do not look for thy father, forget his voice! Write how this city refuses to read the work that answers to his name. Write how this port is afraid of what moves beyond its compass — oh! like an eastern mystery — and is bent on withdrawing once and for all, in its total strangeness, from the city's imaginary control! And above all, write how this childish yet understandable fear is being misused. Write how too many little dupes in these conflict-ridden times are being made a monkey of and led up the garden path. *Ecrasons l'infâme*, Telleke! Depart now, and take the tram ...

Lo! here is a writer, Telleke, ink runs through his veins and his fingers are itching. There it is, hideously ringing its bell, the tram. An alarm clock goes off, it is five o'clock. 'Five o'clock,' writes the writer Telleke, 'I'm leaving this bed.' At that very moment the veil of the Central Station is rent: the express train coming from Paris, capital of the nineteenth century, has bored its way through the façade with the power of many horses. In the Zoological Gardens reigns a merry lunacy: someone has opened all the cages. The garden is transformed into a formless tangle of animals running to and fro, a mad animal circus. Species can no longer be distinguished, there are only differences. This is the beginning of another age.

*Translated from the Dutch by Bart Eeckhout (with special thanks to the author and to Jim O'Driscoll)*

## Notes

[1] By Paul van Ostaijen (1896–1928), Antwerp poet and writer of grotesque stories. A major literary figure undeservedly little known outside Flanders and the Netherlands. [Translator's note] See 'The man with the swine's head', p. 164.

[2] Sinjoor (masc. pl. Sinjoren) ([Sp] señor [gentleman]): popular sobriquet for an inhabitant of Antwerp sometimes used in Flanders at large but more usually sported by Antwerp people themselves. [Collective note]

[3] Other sources make mention of an Arabian oryx and Lesbian pheasants. The chronicler Revius even mentions a small Persian ass, but that is foul gossip.

[4] Name of the main shopping street in Antwerp. [Translator's note]

[5] Commonly used poetical description of the tomb where, since 8 November 1952, Paul van Ostaijen, lies. The Prince's Grave is to be found in the Antwerp cemetery of honour, Schoonselhof. The tombstone is graced by a monument (by Oscar Jespers) representing a man lending his ear. Oral tradition would have it that this man's name is Telleke. [Collective note]

# Pense-bête

*Marcel Broodthaers*

*Ars Poetica*

A taste for secrecy, hermetic practices, for me they're all one, and a favourite game. But I mean to cast shame aside and reveal my sources of inspiration.

Works of jurisprudence have often excited my imagination. Each word in them has its place, a very precise place. The ambiguities of law surely derive from differing interpretations of the text, from the spirit and not from the letter.

Words in statutes shine like diamonds. So there you have it, a passion of mine since I first learned to read. A dangerous passion, an obsession, whose meagre results you have here, a few poems deflected from their natural state, from people and from things.

Here is a selection from my bedtime reading:

*Pork*

Precious morsels
chops hams head
porcelain collar
ears on the snout of the counter ...
I read in your tiny eyes a children's book.

34

## The Parrot

Reinforcements are sent for.
They open fire. He responds with an attack of fog.
Already he screams from another world.
He inflates himself. He will pass on.
(He repeats *vive la liberté*)


## The Cockroach and the Boa

At last I see through myself. I'm afraid of being seen.

I am a boa,
it's the most terrible thing that can happen to a snake.


## The Little Finger

Stemware.
Turning over. Clowning around.

The Ox.
Tightrope walker on a tongue. And good at it.

Maggots.
They subdivide. And never stop.

The Lizard.
I want to invent amazement, he says, and vanishes with his idea.

*The Index Finger*

The house.
Where Captain Courageous hollows his abyss, the house holes up.

The dogs.
My master had a remarkable one. A real blind dog.

Water.
Everything it embraces is smaller than itself.


*The Mussel*

This clever thing has avoided society's mould.
She's cast herself in her very own.
Other look-alikes share with her the anti-sea.
She's perfect.*


*The Jellyfish*

It's perfect
No mould
Nothing but body

Pomegranate set in sand.
Kiss of lips unspoiled.
Bride. Always a bride, in dazzling terms.
Crystal of scorn, of great price at last, gob of spit, wave, wavering.

*Translated by Michael Compton. Broodthaers's play on the difference between la moule (mussel) and le moule (mould) is lost in English. The remainder is translated from the French by Paul Schmidt.

# To the realm of fables:
# the animal fables from
# Mesopotamia to Disneyland

## *Marc Holthof*

In 1927 cartoon film-maker Walt Disney left Hollywood for New York because of a dispute over 'Oswald the Rabbit', a cartoon character with which he had had a modest success. The film's distributor, Mintz, had engaged a number of its cartoon artists, stealing them behind Disney's back so as to make the Oswald films himself. Walt was deeply disappointed, but appeared to be helpless: the rights to Oswald belonged, as was the practice in the film industry in those days, to Mintz. Disney had to return empty-handed to LA: with no artists, and no cartoon character. He had nothing left, his career was over.

Richard Schickel relates in his (unofficial) biography of Disney, *The Disney Version*, that Walt (or the publicity department of the Disney studio) put out the following story in 1934 for the *Windsor Magazine*:

'But was I down in the dumps?' Walt asked himself (in the train going home): '— not in the least. In my heart of hearts I was happy, because out of the difficulties and confusion there grew a cheeky, comic character — at first vague and uncertain — but it grew and grew. Finally, a mouse was born: a frisky, playful little mouse. The idea gradually took me over completely. The wheels turned to its rhythm — it was as if the train was saying "chug, chug, chug, chug, mouse!" The train's whistle screamed it: "A m-m-mowa-ouse!" Once the train had reached the Midwest, I had already dressed my dream mouse in a pair of basic red shorts with two large mother-of-pearl buttons, thought out my first scenario, and was once again on top!' Mickey Mouse, the most famous animal of the twentieth century, was born.


## The human animal

'If you look at an animal closely, you get the feeling that there's a person inside laughing at you.' (Elias Canetti)

Philosophers have since the beginning of time tried to convince us that we are different from animals. We would agree that that is chauvinistic, but in practice we might not be so sure: 'The Serbs are beasts, but at the same time they are men', declared a Muslim fighter in Sarajevo. At the end of the nineteenth century there was an extensive family of Indians on exhibition in the 'Jardin d'Acclimation' — a section of the Paris Zoo — together with their totem poles and wigwams: something which was copied in Antwerp in 1885 with 'negroes' from the Congo and in 1944 with another sort of 'black'.

   People can ask themselves: 'If Man is really so different from an animal — if he really is that reasoning, self-conscious, political, talking, laughing, disciplined being as the philosophers would have us believe — why then does he need all those limiting definitions to prove his identity, to differentiate him from the animals?' Levi-Strauss notes at the beginning of his work *Totemism Today* [*Le Totémisme Aujourd'hui*] that modern anthropology was created out of *fear* of the animal in Man:

To preserve the integrity and at the same time to give a foundation to the mentality of the normal, white, adult man, there was nothing easier than to group together all the practices and beliefs which seemed strange to him — even if they were all different and plucked out of context. An inert mass of ideas crystallized around this, which would have been less innocent if men had acknowledged their presence and activity in all cultures and civilizations, including our own. Totemism is primarily a projection outside our world, through a sort of exorcism, of mental attitudes which cannot be reconciled with the requirement for there to be a division between Man and Nature, which in Christian thinking [and not only there — author's note] is held to be essential for it. [sic]

   Late nineteenth-century anthropology tried to exorcise the animal-in-man scientifically. Lévi-Strauss compares this to the way in which Charcot dismissed

*38*

hysteria (and, we might also add, the way in which Durkheim dismissed religion and the sacred from sociology). All these positivistic sciences tried to get rid of the animal, desire, the sacred — the 'irrational' — from human nature. Under the pretext of scientific objectivity, scientists wanted to make madness, desire, the primitive and religion more different than they really are. This can be explained by the common tendency of different scientific disciplines at the end of the nineteenth century 'to describe human phenomena as each being something separate — and in the form of what I would call a "natural situation" — which the scientists preferred to keep outside their moral universe with the purpose of preserving the clear conscience they cherished against those phenomena'. (Lévi-Strauss) [sic]

The last kind of argument in which the relationship between Man and the animals is explicitly taken into account, where the so-called 'normal, adult, human, white' advantages have no play, is probably the animal fable. In the fable there is still something left over from the metamorphosis between man and animal which Canetti in his *Mass and Power* described:

> [...] the mythical ancestors of the Australian Aborigines are both human and animal, often even human and plant. [...] they are accepted as dwellers of a mythical prehistory, a time when metamorphosis was a general endowment in creatures and took place constantly. The fluidity from that world is often emphasised. A man could change himself into anything, and also had the power to change others.

In the fable people are also animals, if only because the fable is older than the philosophy that men would like to make so unique. Sometimes in ancient 'animal' fables people appear, but not a single animal. This is not a contradiction: Man is an animal just like any other.

### The animal fable

The fable is a short story in prose or verse that shows 'animals' behaving and speaking like humans and contains an implicit or explicit moral lesson. The origin of this genre lies in Mesopotamia, whence it spread on the one hand to India and on the other, to Asia-Minor and Greece. In Greece the fable is inextricably linked with the name of Aesop, a Thracian or Phrygian slave, perhaps mythical, though probably an historical figure who lived in the sixth century before Christ.

When the lion was about to attack the defenceless goat, the goat said: 'Let me go and I will give you my companion the ewe when I get back to the pen.'

'If I let you go, then tell me your name,' said the lion. The goat answered the lion: 'Do you not know my name? My name is "I shall make you wise".'

When the lion came to at the pen, he said: 'I set you free.'

But the goat answered from the other side of the fence: 'And you have become wise in exchange for all the goats who do not live here!' (Old Sumerian fable from the city of Nippur)

There are many texts in the fable genre and they appear over a long period of history, from Mesopotamia to the present day. It is one of the few genres to survive the political and literary upheavals of the early middle ages. Man still uses the animal mirror-image to think about himself. But the lack of differentiation, the interchangeability between man and animal inherent in the genre, made the fable dangerous to Western rationalism. To ward off that threat, the function of the fable was completely changed in the eighteenth and nineteenth centuries. A pedagogic-moral instrument was fashioned out of a cynical worldly wisdom about men–as–animals. The fable crept into a growing number of books (and schoolbooks) and anthologies for children, and became part of the teaching process. The underlying intention was clear: to banish the animal, the passions (and the sacred too?) from human nature. The fable became boring and uninteresting for adults.

The victory of rationalism seems complete today: apart from a few enduring exceptions there is but little interest in animal fables. Their moralizing tone disturbs us — even in the ancient Greek fables which were not intended to be like that. The end of the genre seems close at hand — or is it? After all, is the rationalistic façade anything more than a thin veneer? Has the fable not become so all-pervading in our culture that we are overlooking it?

## The changing function of the fable

Fables surviving from ancient times do not often sit comfortably with the now fashionable ethical/didactic views going back to the theory of Lessing. The ancient fables of Aesop were not precepts on how we should behave, but guidelines for learning to see life as it really is, in all its mediocrity and ugliness. In an amusing way the fables give us a deep, sometimes bitter though often succinct insight into the human soul:

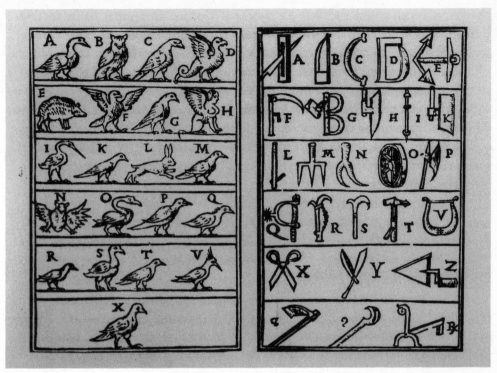

*Animal and visual alphabet, from Johannes Romberck,* Congestorium artificiose memorie *(Venetian edition, 1533)*

The Good Spirits and the Bad Spirits
The bad spirits who wanted to take advantage of the weakness of the good ones, drove them away. The good spirits flew to heaven and there they asked Zeus how they should conduct themselves among men. The god replied that they should not all appear to men at the same time, but one after the other. That is the reason why the bad spirits, who live close to people, torment them unceasingly — while the good ones, who have to come from heaven, only pay them visits with long intervals in between.

The meaning is that we have to wait for the good, but each one of us has to struggle with the bad in our daily lives. (Aesop)

## The fable as a rhetorical figure
Animal fables with a cynical or moral lesson are distinguished by a double

41

rhetorical mechanism: the human world is clothed in animal form so that a human (moral) lesson can be drawn from that animal clothing. This feedback makes man and animal interchangeable terms. In ancient times the fable was often used as an exercise in rhetoric: (student) orators were asked to compose a story on a given subject or to think up a new moral from an existing fable. Aristotle writes: 'Fables are suitable for orations since they have the particular advantage that finding appropriate facts in the past is difficult, while thinking up a fable is easy.' Thus for him too the fable is not a separate genre, but one of the many rhetorical media with which the speaker could persuade his audienc

Nojgaard writes:

This way of looking at the fable was to predominate until in the eighteenth century... when the fable no longer belonged to real literature... where it only appeared in an adapted form — i.e., as part of a reasoning process in which it functions as an example. (...) throughout antiquity the fable was to be used in that way — in all genres, from the didactic poem and philosophical treatise to lyrical poetry, tragedy and comedy.

That the ancient fable was used to *persuade* and not to moralize shows that it was the expression of a communicative, democratic culture — not (or not yet) of the monologue and enforced rules such as we see later in La Fontaine and Lessing.

Rhetoric is in essence a republican art: men must be able to tolerate the strangest opinions and insights — yes, and even draw a certain satisfaction in refuting them. Men must also learn to listen as well as speak, and as listeners men must learn to evaluate the (rhetorical) devices being used. Education in classical antiquity usually culminated in Rhetoric: that was the highest intellectual activity for a well-educated political person — for us a really surprising thought! [sic]

so wrote a young classicist, Friedrich Nietzsche.

### Writing
The fable underwent its first important change of function when it was written down. By committing the fable to paper, this primarily oral genre lost its rhetorical function. It no longer served as an example in oratory to prove the validity or otherwise of a particular topic, but gained its own, independent significance. Because it told the story itself, it was now set to verse and made into a piece of 'literature' through poetic treatment. The precise moment when the fable

became an independent literary genre can be given exactly: the first century AD, when Phaedrus had already set the material collected by Aesop to verse, while the rhetorician Quintilianus continued to classify fables among rhetorical arguments to recommend them as examples for orators who wanted to work on the minds of 'farmers and illiterates'.

More so than Phaedrus, Jean de la Fontaine (1621-95) fashioned the fable into a small literary art-form. 'He was successful in making the fable into an elegant, poetic plaything' noted Lessing. La Fontaine dedicated his first collection of fables to the six year-old Dauphin. The success of his fables therefore was also due in part to the circles for whom he had written them: the French Court. La Fontaine's fables played a not inconsiderable role in the construction of the (sometimes frivolous) lifestyle, morals and customs of what Norbert Elias has called modern court society.

On the order of Zeus, Prometheus created the people and animals. As he observed that the animals were more numerous, Zeus gave him the order to reduce their number and change some of them into people. Prometheus did as he was commanded, but the result is that those who were not originally created as people certainly have human form, but still to this day have an animal's brain. (Aesop)

It is no accident that the fable is the literary genre most influenced by the rationalistic and moralistic *zeitgeist* of the eighteenth century. Above all, it was German rationalism as represented by Wolff, Gellert and Lessing that influenced the genre both practically and theoretically — much more so than the French successors to La Fontaine such as Bellegarde, Du Cerceau, La Motte, Le Brun, Pitaval, J.-B. Rousseau or du Ruisseau. The 'moral upbringing of humanity' (also the title of a work by Lessing) was a central feature of the German viewpoint. For Wolff in 1738 it was a matter of 'making inroads into human stupidity'. Two years later the Swiss Breitunger wrote: 'The fable writer is a teacher of morals; his most important aim is the edification and improvement of humanity'. For Gellert the pedagogic function of the fable is important; he wants to educate intellectuals, youth and plebs, *and* the female sex, in accordance with 'nature', 'virtuousness' and 'dignified' behaviour.

Lessing writes in his 'Von dem Wesen der Fabel' (1759): 'The purpose of the fable, the reason it was invented, is to teach morals.' With Lessing, however, the fable is more than pedagogy: 'If we apply a general moral principle to a particular case, materialize that case and make a story out of it in which the general principle is clearly acknowledged, then that story becomes a fable.' The fable is thus the material realization of a moral doctrine. Lessing's fable-moral does not limit itself to the education of the bourgeoisie. For him it is a matter of the moral nature of men, of humanity at large. Lessing's fable-moral has to be understood in the Kantian sense as moral criticism, not just as moral teaching.

The Ass and the Wolf
An ass came upon a hungry wolf. 'Have pity on me,' said the ass, trembling with fear: 'I am a poor, sick animal. Just look what a thorn I have in my hoof!'

'Truly, I do pity you,' answered the wolf, 'and my conscience tells me that I must put you out of your misery.'

Hardly had he spoken, when the ass was torn to pieces.
(Lessing, *Fabeln, Erstes Buch,* 1759, XXVIII)

Lessing thus recognizes the existence of (and writes) fables in which Man's free-will, and thus pedagogy, plays no part. In these fables a more generally valid ('natural') principle is advanced — for example, the law of the jungle: marten eats hare, fox eats marten, wolf eats fox. For Lessing, cynical fables exist alongside moralistic ones. In addition to fables in which human free-will is allotted a role, there are fables where 'the laws of nature' are allowed to rule. As well as fables that say: 'so must it be', there are others that simply say: 'so it is'.

Johann Gottfried Herder wrote in his *Fabel* (1801): 'Since Aphtonius men have classified fables as wise, moral and mixed.' For Herder, his fables are primarily wise, they are the 'Schoolbook of Nature', they set out the laws of nature. He had some doubts about moralizing fables: 'How can men learn from animals and trees? And from which animals? From the wolf? From the fox? From the marten?'

Herder's opinion was shared by Hegel. In *Die Fabel* (1835) he writes:' ... generally speaking, it is not even of much benefit to put animals in the place of men, because the animal form always remains a mask, which clouds understanding of the meaning rather than clarifying it ...' Here Hegel is implicitly criticizing Lessing. What for Lessing is a necessity for making a general moral law clear — the materialization of the moral in the form of an animal fable — is for Hegel unclear: the materialization covers up as much as it reveals. Hegel's criticism anticipates the collapse of the success of the literary fable in the twentieth century. More even than Lessing, he believes in a clear distinction between man and animal: even a moral animal mask is unsuited to humans. The fable just did not suit Hegel.

## The fable image

If we are to continue the history of the fable further, we must go back in time — in particular to antiquity and the Renaissance where there was a close relationship between the fable and the image. Quintilianus wrote that an orator must make things graphic for the listener ('quae velut in rem praesentem perducere audientes videtur') so that he appears to see more than to hear ('ut cerni potius videatur quam audiri') as if he were present himself ('non aliter quam si rebus ipsis intersimus'). Attention to the illustrative character of the language and, vice-versa, the linguistic character of the image, was above all to be cultivated in the Renaissance under the motto attributed to Horace, 'Ut pictura poesis': poetry should be like painting.

In this tradition, image, fable and moral all belong together. 'In antiquity the fable is defined as an invented story which illustrates the truth with imagery': "logos pseudeis eikonikoon aleitheian". 'From the metaphorical image to the illustration is just a short step. The allegories of the fables stick closely to their imagery, and with the images the stories spread far and wide. A portrayal of the fable of the fox and the stork can be found on a Greek vase from the third-fourth century BC and on a Roman gravestone from the first century AD 'There is no doubt that, from antiquity to the middle ages and even later, there existed an unbroken tradition in which the fable appeared together with the image and the word', writes Tiemann.

In 1419 the Florentine Cristoforo Buondelmonti bought the manuscript of the *Horapollinis Niloi Hieroglyphica* on the Greek island of Andros. This allows us to see how Hellenic work was able to represent abstract ideas (time, for example) in images, particularly as animals (e.g. the eagle). The hieroglyphic tradition of the Renaissance was deeply influenced by the re-discovery of the *Horappollo*. For Guazzo in his *Dialoghi Piacevoli* there was a close relationship between hieroglyphs and fables: 'Con simile artificio, e misterio [as in the hieroglyphs] ci diede Esopo molti precetti inuolti nelle fauolle de diuersi animali, onde si traggono sentimenti morali, e gioueuoli alla uita nostra.' In 1505 the Venetian master-printer Aldus

Ick vont onlanx mijn Lief int groene sitten slapen,
Ick sagh haer rooden mont, ick bleefer op staen gapen:
Dies creech ick stelens lust, ô selsaem dievery!
Ick stal van haer een kus, sy stal een hart van my.
5  Als t'muysken raeckt aen t'speck, het eet met groot verlangen,
Het vat, en t'wert ghevat; het vangt, en t'wert gevanghen:
Wat vreemder streeck is dit! wat rancken can mijn Lief!
Sy sit gherust en slaept, en noch besteelts' een dief.

*'Die Steelt, die pueelt', from Joseph Cats,* Silenus Alcibiades sive Proteus *(Middelburg, 1618)*

46

Manutius issued Horapollo's *Hieroglyphica* under the same cover as Aesop's *Fables*. Images of animals were often used in rhetoric as mnemonics:

particularly informative in this connection is the use of symbolic images in the 'memoria rerum', the art of memorizing. Here, animals were often used in place of abstract ideas, where people were based on the characters learned from the fables: the lion for strength, the hare for fear, the lamb for innocence, and so on. (...) the idea that sight was the strongest of all the senses, was the foundation stone for teaching by memorization which began with Simonides. The pictura-poesis theory depended to a great extent on the same understanding: the poet too leans on images linked to subjects he uses in his work, while the artist paints them on his canvas

writes Tiemann. This movement, systematized by Cesare Ripa among others ('Iconologia'), ends up in the *emblemata* literature, of which Joseph Cats was the most prominent representative. In the preamble to his *Silenus Alcibiades sive Proteus* (1618) we read:

If someone were to ask me what emblemata really were, I would answer him that they are silent pictures, which nonetheless speak: small matters which are nonetheless weighty: ridiculous things which are nonetheless not without wisdom: in which men point out [the way to] virtue as with fingers, and can touch with their hands, into which (say I) men as a rule always read more than there is: and think more than there is to see

This allegorical fable-marriage of word and image finally comes to an end in the Enlightenment, not in the least through Lessing's doing.

In his *Laokoon oder über die Grenzen der Malerei und Poesie* (1766), Lessing writes that the changes an object undergoes are actually the poet's subject, while its visible form are the painter's subject. Thus literature is the art of time, painting the art of space. His judgment on anything that does not conform to this scheme:

In poetry a longing for description is created and in painting a predilection for allegory, with the purpose of making a babbling picture from the one without a real understanding of what can and must be painted, and from the other a silent poem without taking into account the extent to which painting can express universal ideas without abandoning its own metier and degenerating into an arbitrary sort of writing. (...) the attempt to express universal ideas in visual form produces only grotesque forms of allegory.

La Fontaine had already written earlier: 'Les mots et les couleurs ne sont choses pareilles;/Ni les yeux ne sont les oreilles ('Words and colours are not the same;/ Just as eyes are not ears').'

This attitude had considerable influence. Painting and poetry are still considered separate entities and any form of allegorical or illustrative language is disapproved of. Lessing's distinction between the art of space and the art of time is mainly criticized because of its Cartesian dualism: painting refers to the animal and the female body, literature to mankind, the masculine soul. In the words of William Blake: 'Time is a Man, Space is a Woman.'

In *The Marble Faun*, the American writer Nathaniel Hawthorne let the sculptor Kenyon conclude from Lessing's dichotomy that 'in everything that is sculpted there must be a moral standstill, given that of necessity there is also a physical one.' W.T.J. Mitchell finds traces in Lessing's *Laokoon* of what he calls the rhetoric of iconoclasm:

a rhetoric of exclusion and domination, in which a caricature is made of others like someone with irrational, obscene behaviour in which we (fortunately) have no part. The images of the idol worshippers are typically phallic ... and must therefore be castrated, feminized, their tongues must be cut out to deny them the power of expression and persuasion. they must be made 'dumb', 'stupid', 'empty' or 'illusory'.

In short: the animal must be driven out.

Lessing's dismissal of the illustrative character of language goes hand in hand with his views on the fable. The 'real' tolerates no decorations, embroidered verses or illustrative examples. It does not depend on sensuousness, but on the rational argument of a moral. 'A fable is bad and absolutely does not deserve the name of fable, if the deed it tells is allowed to be portrayed in its entirety. It then contains only an image, and its painter has not fashioned a fable but an emblem.'

Lessing's rational reduction of the fable to a moral literary genre unintentionally heralds its end. Without the interchangeability of man and animal, the genre is doomed to disappear. The subtle distinction that he himself makes between natural and human morals, between cynical and moral fables, was not recognized by his contemporaries. They were to fashion the fable into a mere pedagogical instrument. Gellert for example legitimizes and popularizes what was in essence already present in La Fontaine: the animal fables become free and uncommitted, ironically funny, but in their critical sharpness dull little stories in verse. At this point the most important condition is fulfilled for the genre to fall into the hands of the pedagogue: the form is watered down, the characters and situations of the classical fable lose their sharp edges in the cause of still greater emphasis. Their sharp inventiveness was filed down into a moral sermon, repetitions replace humour and persuasive power; the fable becomes a story of 'example' with little characterization. In the cause of pedagogy the genre develops into a piece of trivial literature for the consumption of the masses. Inroads were already being made at the beginning of the 1770s thanks to the great success of the first weeklies for children.

Pastor Johann Andreas Christian Löhr ends a fable from his *Fabelbuch für Kindheit und Jugend* (1816) with: 'Now be good, suffer and be tolerant my dear child! Things will blow over, and certainly more easily than with pride and unruliness.' [sic]

Need we say that here the worldly wisdom of an Aesopic fable is being completely turned on its head?

## A new fable medium

At the beginning of this century the popularity of the fable was going rapidly downhill. Only a few attempts, usually highly individual (Kafka, Brecht, van Ostaijen, Thurber) were made to instil new life into the fable. And they have more to do with the fable of antiquity than with the rationalistic-pedagogic one from the eighteenth and nineteenth centuries.

However, we do find the fable (both moralistic and cynical) at the beginning of the twentieth century in a new medium: the cartoon film. This was a new medium that was perfectly suited to the illustrative character of the fable in antiquity and the Renaissance. The film breaks down the barriers between visual art and literature that Lessing in his *Laokoon* had drawn up: film is always both pictorial and narrative, an art of space *and* time, it is mechanically reproduced 'ut pictura poesis'.

Starting with Winsor McCay and his dinosaur *Gertie* (1914), the animation film teemed with anthropomorphic animals and all sorts of cynical, moral wisdom. The relationship to the ancient fable becomes very explicit in 1920. Producer Paul Terry received a phone call one evening from a young director, Howard Estabrook, who told him that he had an idea for a series of animated cartoons based on ... Aesop's fables. Terry asserted later that he had never heard of Aesop, but still saw something in the suggestion. 'Of course, once they were ready, they didn't much look like Aesop's fables any more,' he explained, 'but it was a start.'

With support from the Keith-Albee Vaudeville circuit, Paul Terry produced his *Aesop's Fables: Sugar-Coated Pills of Wisdom* at the rate of one a week from 1921 to 1929 — they were short, silent, black and white cartoons packed with animals. With their sugary wisdom, they turned out better than expected, however, because in the cartoons cynical, healthy common-sense gets the better of the moralizing.

Terry's *Fables* would have been long forgotten if Walt Disney had not taken them as his great example: 'Even back in 1939 my great ambition was to make cartoon films as good as the *Aesop's Fables* series', declared the master of the cartoon film. And Leonard Maltin adds: 'Disney later explained that Paul Terry's *Aesop's Fables* formed the basis for his characters at that time.'

A considerable part of Disney's films hark back to the moralistic/pedagogic tradition of the Enlightenment. It is remarkable how in the later Disney films (mainly) all the animal aspects (even in 'realistic' nature documentaries) are suppressed for the benefit of sentimental anthropomorphosis. Elsewhere in the cartoon the interchangeability between people and animals is preserved. If an animal in a Tex Avery cartoon loses part of his fur through some act of violence (explosion, circular saw, etc.), he appears wearing human underwear underneath ...

The great strength of the medium of cartoon is its ability to illustrate everything perfectly, to connect everything with everything else, for everything to flow over into everything else. The cartoon film is the modern equivalent of the dream-world of Canetti's aborigines. Pioneer Max Fleisher often uses metamorphoses in his Koko films from the 1920s. So in *Koko the Kop* we see him chasing the dog Fitz, who suddenly jumps up on a wall and changes into a window, complete with an attractive housewife inside. The lady flirts with Koko, he kisses her and just at that moment lady and window turn back into the dog ...

*How reality finally becomes a fable* (*Nietzsche*)

With his famous cartoon and animal films — which helped to shape the consciousness of many generations — Disney built up an empire. He constantly referred not only to H.C. Andersen and the Brothers Grimm as his great examples, but also to Aesop's fables (seen of course through rationalist/moralist spectacles). Many years later, in 1955, Walt realized an old utopian dream: Disneyland. His aims went much further here than just setting up a sort of fairground: he was fed up with the wild urbanization of Los Angeles and created Disneyland as an alternative, as an example.

Max Weber pointed out that capitalism was not entirely unethical. On the contrary, it is precisely the result, the realization of the Protestant middle-class ethic. With his 'Main Street USA', his 'Liberty Square', his 'Frontier-', 'Adventure-' and 'Discoveryland', Disneyland is the realization of the utopian ethic of capitalism. Louis Marin shows in his 'Jeux-Utopiques' that Disneyland was conceived and structured as a capitalist myth.

There is a little film in which Disney appears leaning up against the model table of EPCOT (Experimental Prototype Community of Tomorrow), a city (not a pleasure park) of the future, while he explains his ideas and his dream. His life's aim, he declares, is to build this ideal city from which all violence would be banished, in which conflicts between social classes would be got rid of ... The same 'enlightened' ethic is also to be found at the heart of Disneyland, Disneyworld and EuroDisney. The kingdom of Mickey Mouse is in fact the ultimate materialization of the moral dream of rationalism. Disneyland and EuroDisney are moral fables made real, they are colossal materializations of the general capitalist ethic — entirely in accordance with Lessing's prescription. Thanks to Walt Disney, we can now make a trip into the kingdom of fables where we can experience materially the rationalistic lesson in virtue — all packaged up as the 'American Dream'.

Certainly, the form in which this has all been packaged would make the hair of a rationalist like Lessing stand on end. Disneyland is a monster of oratory, an anti-Laokoon. Lessing's beloved fable moral is conveyed here purely through sensestimulation, purely illustratively, full of adornments — like a gigantic Renaissance allegory, a modern, technological version of the Bomarzo's monster garden.

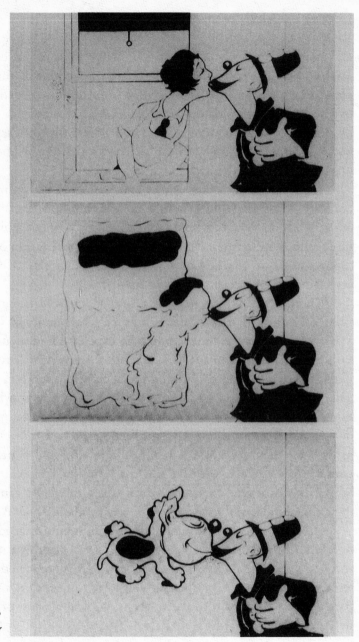

Koko the Kop
*by Max Fleisher*

*'Sugar-coated pills of wisdom' by Paul Terry, 1921-9, from* Aesop's Fables

Disneyland shows that the moralistic animal fable is still with us in the twentieth century, and indeed forms the central metaphor, the backbone of late-capitalism. 'Disneyland exists to conceal the fact that America is Disneyland', wrote Jean Baudrillard. We might also add: Mickey Mouse exists to conceal the fact that Man is a mouse.

*Translated from the Dutch by Milt Papatheophanes*

## Bibliography

Aristotle, *Rhetoric*, vol. 2 (bk II) — French version. Text compiled and translated by Médéric Dufour, collection of the Universities of France, published under the patronage of the Association Guillaume Budé, Paris 1960

Elias Canetti, *Masse und Macht* (Düsseldorf 1960).

Emile Chambry, (ed.), *Esope-Fables,* collection of the Universities of France, published under the patronage of the Association Guillaume Budé (Paris 1960).

Herbert Kaiser, 'Die Pädagogisierung der Fabel am Ende des 18. und zu Beginn des 19. Jahrhunderts', in *Die Fabel — Theorie, Geschichte und Rezeption einer Gattung*, Peter Hasubek (ed.), (Berlin 1982), pp. 163-279.

Leibfried Erwin & Josef M. Werle, *Texte zur Theorie der Fabel* (Stuttgart: J.B. Metzlersche Verlagsbuchhandlung 1978).

Gotthold Lessing, Ephraim *Werke* (Munich 1970).

Claude Lévi-Strauss, *Le Totémisme aujourd'hui* (Paris 1962).

Leonard Maltin, *Of Mice and Magic — A history of American Animated Cartoons* (New York 1980).

W.J.T. Mitchell, *Iconology — Image, Text, Ideology* (Chicago & London 1986).

Friedrich Nietzsche, *Rhetoric and Language* (French version). Text translated, introduced and annotated by Philippe Lacoue-Labarthe and Jean-Luc Nancy in *Poétique 2* (1971), pp. 99-141.

Morten Nojgaard, *La Fable Antique* (2 vols, Copenhagen 1964 & 1967).

Karl-August Ott, 'La Fontaine als Vorbild — Einflüsse französischer Fabeldichtung auf die deutschen Fabeldichter des 18. Jahrhunderts' in: *Die Fabel — Theorie, Geschichte und Rezeption einer Gattung,* Peter Hasubek (ed.), (Berlin 1982) pp. 163-79.

Barbara Tiemann, *Fabel und Emblem. Gilles Corrozet und die französischer Renaissance-Fabel,* (Munich 1974).

# Modern, postmodern and archaic animals

## Stephen R.L.Clark

Common sense assures us that many non-human animals have desires, thoughts and plans much like our own. But common sense can be challenged. The chief principle of a properly scientific Enlightenment is that nothing is to be believed merely because 'common sense' would have it so, or because we *feel* its truth. Only what can be proved to the satisfaction of someone determined not to practise a potentially misleading empathy can be trusted: animals must be treated 'objectively', for the real world is one of mere, indifferent objects unaffected by our casual or conventional likes, sympathies and identifications. This 'modernism' is itself under challenge from 'postmoderns', inclined to doubt that there are any truths at all outside the sphere of what 'we' happily endorse: 'the objective world', by that account, is only another fanciful creation, and those who insist that we must only think 'objectively' are pretending to a mystical insight into 'reality itself' that postmoderns 'know' (or choose to say) is quite impossible.

It follows that any sensible examination of the problems posed by an enquiry into the consciousness, or otherwise, of animals (which is to say, non-human animals) must begin by considering the Enlightenment project that has given us modernity and, more recently, postmodernism. That project is founded on the twin postulates that knowledge is only possible if we divest ourselves of our human emotional nature, and that what is really knowable has no pre-scriptive force. It is founded, in fact, on the rejection of moral objectivism. I have argued on other occasions that a satisfactory solution to the question whether, and how, animals are conscious can only come by a return to moral objectivism, to the doctrine that knowledge arises from a loving attention to what is knowable, and that what is known makes its own demands on those who know. I have also sought to identify the clues and cues that give loving observers their understanding of the non-verbal. In this paper I shall instead consider what animals are for modern, postmodern and archaic thought, with-out directly addressing those other, wider issues.

## Modernism

Modernism begins in the conviction that what matters for an understanding of the world, and its proper treatment, are those properties that are independent of human discourse. One man may be, conventionally, 'of gentle birth', and another 'of base stock', but this difference cannot be a good reason to treat them differently. Instead we should have regard to their God-given natures, what they are in origin. Merely conventional moralists of course may insist on these conventional distinctions, but what matters so far as the objective moralist is concerned must be what is true apart from such convention. Obviously enough: if it could be right to enslave one and wrong to enslave another merely because the one was *described* differently from the other, then that moral judg-ment at least would be true only in virtue of what 'we' said, and therefore not be objective. Objectively valid judgments can be grounded only on what is objectively true. Justice is the moral entitlement of anyone, no matter where she lives or what her accidental caste may be. Natural equals should be treated equally. Factual judgments must similarly exclude all merely human figments: what is can only be what would be whether or not we were here to misperceive it. As Democritus, Father of the West, declared long ago: only atoms and the void are real, and all the rest convention. From which it eventually follows that moral judgments must be non-objective — and conventional ones therefore as acceptable as any! If there are no 'real', no natural, moral obligations, but all such obligations rest upon 'our' serious preference, then it is no longer easy to insist that only 'natural' divisions count. Where all is convention, merely conventional divisions are as good (or bad) as any. Why should it matter any longer that merely naked humans, stripped of their historical and cultural baggage, are only and entirely human? It takes a moral effort to remember this, and why should we make that effort if the only rules that bind us are our own?

There is a further problem. It may at first seem easy to distinguish the moral (which is conventional) from the merely natural (which carries no prescriptive force). We may conventionally decide that the cassowary is not a bird[2] — that is, is not to be treated like more usual birds — or that pets and pigs are treated differently. Whatever we decide will not be *false*. The facts of the matter, though, are otherwise: whatever it is we say or do, the cassowary *is* a bird (that is, cassowaries are winged things descended from the same common stock as blackbirds and penguins) and pigs need not be naturally different from the creatures we make 'pets'. We may not choose to attend to what pigs feel or fancy, but it will still be true that they feel pain when burnt, as much as pets do. But recent writers have begun to deny that facts are natural, any more than values: 'to attribute feelings to X is only to remind ourselves that it is wrong to hurt X [which is to say, respectable people don't usually approve of doing it]'.[3] Can we still insist that *our* conventional divisions are more rational than those of others?

In the remote past, maybe, people believed that 'weeds' or 'creepy-crawlies' named true natural kinds that were intrinsically evil. Then we began to think that these were very partial judgments, that the things we named as 'weeds' were not intrinsically weeds, nor evil. A true morality sought to discover natural kinds of a less subjective sort, the real divisions between this and that. The dogs we pet and the pigs we keep in pens are not so different (except in conventional value) that we should treat them very differently. Then we abandoned moral objectivity, including — though we did not always notice it — the obligation to treat natural equals equally and think less of all merely conventional discriminations. And finally, some few of us begin to deny even the old truth that there are natural kinds at all, that there are 'real equals' that should — on liberal views — be treated equally. We can no longer maintain the distinction between 'merely conventional' divisions and real ones. On this basis it becomes a sufficient defence of discriminatory practices that 'we' identify the victims thus and so. There are no objective rules of justice stipulating that this is wrong, nor any objective distinction between real and merely conventional distinctions. 'Rationality [in matters moral] is a myth.'[4]

The new anti-realism is in one respect more realistic than the old anti-moralism. Anti-moralists could agree that burning a pig alive would cause it pain, but deny that such an event was intrinsically wrong. Wrongness, they said, was neither a logically necessary corollary of causing-to-die-in-pain (for we could all *understand*, even if not admire, someone who said that acts like that were good), nor yet an identifiable property naturally occasioned by non-moral properties and having its own effects on future history. The 'wrongness', they said, was only 'our' projection. But what can it mean to say that something is in pain, if not that it is in a state worth fleeing from? And how could we decide that something was in pain except by acknowledging its screams, squirms and bloody sweats as pleas to desist? Malebranche heard a yelping dog impassively, as uttering no more than squeaking gears: to hear the yelps as evidence of pain would have been to be moved by sympathy. Ordinary descriptions are both factual and moral in their implications: to hold the moral implication off (with a view, maybe, to contradicting it) is to diminish our understanding even of the fact. What sort of pain is it that is not to be avoided? Can we distinguish between philosophers who deny that dogs feel pain, and ones who deny that we should ever mind? More generally: if what there is can never show us what to do, why trouble about what is? If there are no objective values, what value has the truth?

## Species natures[5]

There is a further difficulty for modernists, that our scientific authorities do not actually believe in natural kinds at all. There *are* at least no natural *biological* kinds in the sense once intended. 'Natural kinds', so-called, are sets of creatures with a shared, distinctive nature, but biological taxa, including species, are not so defined.[6] Even if all members of such a taxon happen to have shared, distinctive properties that is not why they are its members. Cows are not mammals because they feed their young on milk; bovine mothers feed their young on milk (unless prevented) because they are mammals. Being a mammal is being genealogically linked with a complex individual, the order Mammalia, such that its members are more closely related to each other than to members of any other order. This is not to say that they more closely resemble each other. The order's members, or those now judged to have been its members long ago, were not always more closely related to present-day mammals than to their non-mammalian contemporaries. Even now, there may be mammals that *look* more like non-mammals than they look like any other existing mammals. There might even be mammals whose parturient females do not secrete milk, just as there might be birds without wings or feathers. They are not therefore 'imperfect mammals', though such phrases are not wholly unhelpful. When Aristotle identified seals, for example, as 'deformed quadrupeds', he was partly right — though any implication that seals are therefore not what they should be must be resisted (Aristotle also suggested that women were deformed men!).[7] Whereas philosophers still tend to believe that there are 'typical' members of a taxon, and to be as eager as Aristotle was to identify defect or anomaly, modern biologists think that cheetahs are as obviously Cats, Down's Syndrome children as obviously human, as any 'type-specimen'. Either might have *been* the type-specimen of the relevant taxon, because the biological type of a taxon is simply the specimen (however unusual it eventually turns out to be) that serves as the referential tie for that particular taxon.

Biological taxa are individuals.[8] That may seem thoroughly mistaken: surely taxa are sets of individuals who more or less 'resemble' each other? So Quine: a biological kind is the set of all things 'to which [the paradigm] *a* is more similar than *a* is to [the foil] *b*'.[9] There may be such sets, but they are not the same as taxa. *Drosophila pseudoobscura* and *Drosophila persimilis* are sibling species, indistinguishible to naive observers, but certainly distinct (because their members do not successfully breed together). If one such species vanished from the world, but there later appeared creatures indistinguishible even to an expert eye (but having a different ancestry) the older taxon would not have reappeared. The dodo, once extinct, is gone forever, because 'dodos' are not just those creatures that look more or less like the pictures, nor even those creatures whose DNA looks more or less like that of the old birds (if we could discover this). There is no need for members of a given taxon to resemble each other more than any of them do members of another taxon. There is no need even for them to share any particular genes which are not shared with creatures of another kind. So being a member of that taxon is not a matter of instantiating any non-historical property, whether obvious or hidden away. The taxon is an individual, and ordinary individuals are parts of it, segments of a lineage. 'Academic classification extends to classes, which it divides according to resemblances while natural classification divides according to relationships, by taking reproduction into account.'[10]

Where two or more species emerge within a previously existing taxon that till then had only been *one* species it is customary to give both of them new names, even if one such species is indistinguishible to us and to its members from the older species. If the earlier species merely develops, as a single population of interbreeding individuals, taxonomic practice varies: some will judge that the differences are such that if the species had been contemporaneous they wouldn't form a single breeding population; others that there is no clear break between the old and new, which may as well be counted as one species. *Homo erectus*, *Homo habilis* and *Homo sapiens* can be one species or three. Traced back through time, of course, the difference between species will often seem quite arbitrary, even when there are two or more daughter species to consider. Why, after all, should x be a different species from y merely because there is another species, distinct from x but equally descended from y? If that other species had not been discovered, or had been extinguished before it was truly established, x would be uncontroversially the same species as y. Even at one time there are populations which reveal the transience of species: the different varieties of herring gull described by Richard Dawkins are one species in the sense that genes from one variety can spread, by degrees, to any other, but two or more if particular varieties are paired. This phenomenon is, in Kant's terminology, a *Realgattung*, a historical collection of interbreeding populations, and in more modern terms a *Formenkreis*, or ring-species. It is probable that humankind, historically, is such a *Realgattung*; it may even be that it still is, that there are particular varieties within the species that would be judged different species if the intervening varieties were lost.[11] As Dawkins points out, humans and chimpanzees are judged to be of different species precisely because the intervening varieties are indeed extinct.

Once synchronic species barriers are established the flow of genes will be restricted: that is what species barriers are, and that is why lions and tigers are of different species. But their ancestors were not, and genes flowed equally from the Urcat to lion and tiger populations. It is because there are, by hypothesis, no diachronic species barriers, that some have reckoned that palaeospecies (like *Homo habilis*) are only metaphorically species at all. Nor can we always claim that the barriers against interbreeding were established by some other general change in the character and conduct of ur-lions and ur-tigers. More likely the barriers were established, by distance or mountain or river, for reasons having nothing to do with any original characters, and the other general differences accumulated since. We cannot even be sure that what had seemed like barriers against interbreeding are always more than accidents of preference or opportunity: maybe lions and tigers *would* interbreed successfully often enough to identify them as a ring-species if enough of them had the chance. There is good reason, after all, to think that domestic dogs, wolves, and coyotes are really all one, variegated species such that not all its varieties will willingly interbreed. Lose all the other dogs, and wolf-hounds and chihuahuas would be unlikely conspecifics — far more so than Voltaire reckoned 'Hottentots, Negroes and Portuguese'.[12]

But surely human beings are all one species in a more important sense than this? Maybe other grades of taxon, family or order or phylum, are merely genealogically united. Maybe there are taxa that look like species but are really not. Species, real species, are, precisely, special. Don't members of a single species share a nature? Don't human beings? Isn't that the axiom on which humanism and the United Nations charter depend? If modern biological theory suggests that there need be no shared natures, no perfect types, and even that the purity of species must be questioned, so much the worse for biological theory. The Negro and the American *must* be our brothers and sisters, and therefore must be like us. It is insufferable to suggest that there are real varieties of humankind that might not willingly interbreed. Still less sufferable to imply that *Pan*, *Pongo*, *Gorilla* and *Homo*[13] might perhaps have been, or still may be, a ring-species. What varieties of *Homo* might breed, or once have bred, with *Pan*? Isn't there a submerged, and prurient, racism at work here? 'Ape' is an easy, racist

61

insult. Those whites who use it should perhaps be reminded that most commentators, world-wide, will suspect the smelly, hairy Europids (which is, 'the whites') of being 'closest' to the ape. Do we need to abandon biological science to avoid being racists?

Racism, on the contrary, is the natural expression of misplaced essentialism, the belief that groups embody different natures. Or else it is an early version of the barriers against interbreeding that establish distinct species. New species is but old race write large. There need be no antagonism between such new-born species: the consequence may actually be a lessening of competition, because the species eventually graze in different places and on different things. There may also be a reduction in adaptive variation, since one newly distinct lineage has lost the input from the other. There may be many good reasons for us not to allow the emergence of barriers between human varieties, and maybe the other breeds of hominoid are now sufficiently distant (as once they weren't) from 'ours' as to make all present interbreeding doubtful. But we do not *know* this to be true. I do not recommend the experiment — but mostly because the fate of any such cross-breed as that pictured by Dawkins with the aid of a computer-generated image would probably be to serve as laboratory material. As long as misplaced essentialism rules, we will suppose that cross-breeds do not really share our nature, that they are throwbacks to a pre-human kind not 'of our kind'. The truth is otherwise.

A further difficulty for moralists is the rejection of norms in nature. If there is no one way of life and character which best suits all or most members of a particular kind, such that we may detect deformity, disease or deviance by comparison with that ideal type, can there be 'a good human life'? Can we truthfully suggest that battery chickens are deprived by being denied 'the' life that chickens would live 'in nature'? If species are only genealogical groups, such that members need not especially resemble each other, we have no right to suppose that there is one way only (however vaguely defined) for any particular species. The limits of variation will be empirically discoverable: what a kind can adapt to will be shown by what as a matter of fact it does, and there will be nothing normative about its 'natural' life. Lineages evolve to make and fit their environments, or else are extinguished. I am a member of the Clark family: but not because I resemble other Clarks, nor yet because there is a way that Clarks will naturally live that is unlike the way that others live. Even if Clarks were more inbred than they were (and so approximated the condition of a species) they need not always resemble each other. There might be atavisms, sports, changelings, or disabled Clarks, but they would all be Clarks, and such variations would not be failures: on the contrary, they would contain the Clarks' hope of posterity. Variation is not a dysfunction of sexual reproduction (even if animal breeders are annoyed if the line they are concerned about does not 'breed true'): it is what sex is for.

This may seem old news. After all, moral philosophers have insisted for most of this century that no natural facts are norms, that 'natural' is not necessarily a term of praise. They have even insisted that human beings are special because they have no single species-nature. The claim is flawed: partly because nothing has such a nature, and partly because the claim exactly identifies a nature shared, at least in potency, by every human being.[14] As Aristotle said, we are creatures whose life is one of acting out decisions. Aristotle was less essentialist than his heirs, because he never identified that 'we' straightforwardly with a species. Not all those born into our species will be capable of their own actions. In some the capacity for choice, or even for understanding, is missing from the start. The good life for us is the life well-lived by those who ask such questions. The very same moralists who have emphasized our freedom from natural constraint actually think but poorly of those who do not, or even cannot, make their 'own' decisions, and constantly deplore attempts to make divisions within the species, which we ought (by their account) to be entirely free to make. So once again the primary danger is not from the biologically grounded notion of kinds, but from our habitual confusion of species and natural kind. We can find out a lot about what individual creatures need and like, what sort of lives they can arrange to live together. We do not need to think that there are goals that only and all conspecifics share.

A fully Aristotelian ethic can accommodate the remarks on species-natures that modern biologists endorse. 'We' means only those engaged, or potentially engaged, in this sort of conversation. 'We' are probably all human, as being members of the *Realgattung* of humankind. But not all our conspecifics need be

63

self-motivating or rational in any way. 'Maybe all triangles must have three angles, but not all reptiles must have a three-chambered heart, though in point of fact they might.'[15] By the same token, Monboddo was right to think that not all human beings *must* be able to speak. And creatures who are not now of our species, though their ancestors once were, may share as much with us as any 'disabled' human. The thought is a dangerous one, no doubt. We are not long removed from moralists who deployed Aristotle's ethics to suggest that Amerindians were natural slaves, owed no respect as real images of God. Any suggestion that not all our conspecifics share our nature is heard as licence to oppress and kill. But there is no such licence, nor any proper argument from neo-Aristotelian premises to ignore the ties of kindred.

## Animals within the text

The new anti-realism is realistic also in this: our actual, lived worlds are struc-tured by convention. 'Weeds', 'creepy-crawly things', 'pests', 'pets' and 'sacred cows' and 'people' are all terms at once strongly relativistic ('Everything green that grew out of the mould/ Was an excellent herb to our fathers of old'[16] — but not to the average suburban gardener) and strongly prescriptive (they carry their recommendations on their faces). It is always a slight shock to realize that other peoples think us odd or filthy for those practices that seem entirely 'nat-ural' to us — but that is merely evidence, for most of us, that *they* are odd or filthy. Older moral realists would often say that other people's discriminations are all superstitious — usually sublimely unaware that the speaker had her own absurdities. Hindus are superstitious for defending sacred cows; Koreans are bes-tial for killing and eating dogs.[17]

The world we live in is full of accidental and historically grounded associ-ations and taxonomies. Churches and council chambers are more than piles of bricks, and more than buildings for crudely commercial purposes. People — even in these liberal days — are also family members and name-bearers, who learn their tasks in life from ceremonials as much as from a would-be systematic course of study. Ancient trees and hedgerows are not replaceable by plastic replicas. Horses and dogs and cows and sheep carry along with them a story dat-ing back to neolithic times, and subtly modified in every generation by the tales we tell. Horses are imagined into being, as much as ridden: cousins to the centaur and the pegasus. Animals inhabit the same story as ourselves.[18]

Individual animals of that kind also have their own biographies. O'Donovan, commenting on Abram's sacrifice to feed three visiting angels, remarks that no-one ever needed to ask 'which calf?'[19] But this is in error: many individual animals are and have been known as such, from the First Cow Audhumla (from Norse stories of the Very Beginning) to the latest champion. When the prophet Nathan confronted David with his sin, it would have been no answer for the King to have said that the poor man's ewe should be replaced: individuals are irreplaceable, even if someone, something else could play their part as well. Animals, like human beings, are identified as individuals in being attended to, in being irreplaceable for good or ill. In that sense even Alexander Beetle is an individual: not that there is or would have been a beetle of that name without a human act of naming, but that — once named and attended to — he is more than 'just an animal', more than a replaceable part. Does that naming make a difference to *him*? Who knows? It makes a difference to dogs and horses.

The realm of human story gives being and significance to landscape, seascape and townscape, to trees and animals and peoples. In the days when there were *real* distinctions to be made, of more importance than the merely fictional, it mattered that — we thought — all humans were alike in being human, all animals in being unthinking brutes. But if that vast distinction is a literary trope, we are released to notice that the world of our 'significant individuals' is populated not only by individual human beings but also by dogs, cats and horses with particular names and values; that the world contains innumerable avatars and images of the Red Bull and the Horse of Heaven. Not all insects are creepy crawlies; some are 'singing masons building roofs of gold', the image of ideal community.

Human beings and human speech are historical inventions as well: our actual experience for long enough was of 'ourselves', the local tribes of people, dogs and horses, and of the 'others', *theria* (wild beasts) and *barbaroi* (who make noises that only vaguely sound like speech, as 'rhubarb, rhubarb'). Slowly we have invented the idea of 'humankind', a universal human essence discoverable over centuries and thousand-miles, distinct from all the other nations of the world, the only 'speaking peoples'. Other ages had no doubt that the other, non-human inhabitants of earth had voices, that their lack of human speech was only a sad disability, on a par with our own ignorance of other human and non-human tongues. The common speech was lost at Babel, but could be recovered.

An appreciation of the historical roots of our present attitudes enriches our present experience of human and non-human neighbours. Attempts to eliminate them on the plea that dogs are only canine, ancestral lands just earth, or spring flowers only 'genital and alimentary organs of plants' were always barbarous.

Don't you see that that dreadful dry light shed on things must at last wither up the moral mysteries as illusions, respect for age, respect for property, and that the sanctity of life will be a superstition? The men in the street are only organisms, with their organs more or less displayed. For such a one there is no longer any terror in the touch of human flesh, nor does he see God watching him out of the eyes of a man.[20]

Even of a fish it is blasphemous to say that it is *only* a fish.[21]

A foolish work of elementary English criticism sought to 'debunk ... a silly piece of writing on horses, where these animals are praised as the 'willing servants' of the early colonists in Australia' (on the plea that horses are not much interested in colonial expansion). C.S. Lewis comments that its actual effect on pupils will have little to do with writing decent prose: 'some pleasure in their own ponies and dogs they will have lost: some incentive to cruelty or neglect they will have received: some pleasure in their own knowingness will have entered their minds' — but 'of Ruksh and Sleipnir and the weeping horses of Achilles and the war-horse in the Book of Job — nay, even of Brer Rabbit and of Peter Rabbit — of man's prehistoric piety to "our brother the ox" they will have learnt nothing'.[22]

If there is no truth-by-nature, or none that we need acknowledge, such reductivism is ridiculous as well as vulgar. Koreans may not be factually wrong to think of dogs as dinner: neither are we wrong to think of them as friends. Which story would we rather tell, which story choose to live in? Better: which story is already telling us? Maybe we ourselves, no less than Black Beauty or White Fang, are characters in a novel?[23] 'We enter human society, that is, with one or more imputed characters — roles into which we have been drafted —

and we have to learn what they are in order to understand how others respond to us and how our responses to them are apt to be construed .... Mythology is the heart of things.'[24] The really important supplementary is: whose is the story? Ours? No-one's? Nobodaddy's? God's? Saga, tragedy or farce?

## The written word

Those who retain any hope at all of realistic understanding of non-human creatures may put their trust in a kind of loving attention rather than the objectivizing curiosity that has eliminated real creatures and real kinds. Loving attention to the creature's particularity (which is as necessary in understanding our conspecifics as in understanding baboons or bees, and sometimes more difficult) gives us hope of discovering the creature's *Merkwelt* and *Umwelt*. The terms are Uexkuell's,[25] and signify the world of cues that are significant for a given animal, and that it notices.[26] Different creatures do genuinely inhabit different worlds, and it takes an effort to identify what things are like *for others*. What is worth emphasizing, and should be noticed by epistemologists, is that not all marks are merely natural. The sheep tick who waits 'patiently' for up to eighteen years until butyric acid triggers its leap from a grass stem to a sheep responds — perhaps — only to fixed stimuli. Perhaps there is no need to imagine that it *senses* anything at all, any more than a photoelectric cell. The chemical causes the muscles to twitch in a way moulded by millions of years of selective pressures. Any awareness of the event might really be epiphenomenal. But some marks — and we know far too little about ticks to know whether this applies to them — are actually created by the animal. Scent-markers create a map embedded in the physical, a set of directions laid down by the animal and by its peers. Derrida, though not for reasons that he would or could endorse, was right to suspect that 'writing' (largely so-called) is older than mere speech. It is by the use of scent-marks and scratches that we create enduring objects. What is it to be the *same* lamb as the ewe has known before? For the ewe, it is to smell the same, to blend with her own smell. So when we, being vocal animals, at last begin to speak of things, we are speaking of our own markers (as well, no doubt, as markers placed 'by nature').

There does seem to be evidence that many animals have a very 'practical' perception of the world, that they are confronted not by continuing objects but by occasions for specific actions, marking out routes through immensity. We suppose that there is a physical universe surrounding us and them, a world where continuously existent objects are available for any purpose of ours but are not defined by those purposes. Male robins can be deceived into attacking any piece of red rag, but fail to respond to models that are — to our eyes — much more like male robins. Robins are not stupid: what is 'enough like' a rival to deserve attacking is not the same for us and them. But perhaps they do not imagine robins, or male robins, at all: there are only occasions in their universe.

But of course the same is true of us. We too inhabit a world marked out for use, composed of things that are judged 'the same' by virtue of their natural and man-made significance. Would an intellectualizing robin be amused to find us responding in 'the same way' to a line of lights, a scrawl of chalk, a string of vocables that all say (so we say) what 'male robins are red-breasted' says? The sameness is so obvious to us that we forget its conventionality. Again: the standard examples, in much modern philosophical discourse, of real, material objects turn out to be artefacts: as tables, chairs and houses. The point is not only that these things are *made*, but that they are perceived as tables, chairs and houses because 'we' choose to use them so. We live in world of *Zeug*, of implements, fenced off from our occasional imagining of a world-in-itself by our wish to complete our projects. Weeds, trees and creepy-crawlies are as conventional. Understanding other creatures' worlds (whether those creatures are of our species or not) is to share, imaginatively, in what they're doing. In so doing we can take some comfort from the biological theories that I have described above: there are no necessary limits on our (and their) capacity for mutual understanding, because there are no natural biological kinds. There are instead only the creatures with whom we co-operate to tell our story.

## Outside the text

So what is it that lies outside our text? One answer, the 'modern' one, is that the 'real world' is the one and only universe revealed to an objective, scientific eye. We must exclude all morally loaded (and so all mentalistic) description if we are to approach the Truth. Such a truth, of course, will have little to do with our ordinary living.

We may if we like, by our reasonings, unwind things back to that black and jointless continuity of space and moving clouds of swarming atoms which science calls the only real world. But all the while the world we feel and live in will be that which our ancestors, and we, by slowly cumulating strokes of choice, have extricated out of this, like sculptors.[27]

Even as an ideal limit the jointless universe is lacking: in what sense, after all, is it ideal? Why should such a truth concern us, and what grounds the devotee's conviction that it is the truth at all? The attempt to think it through consistently must in the end require us to abandon the strange superstition that we *think* at all. Eliminative materialists profess to believe that neither they nor anyone else *believe* anything at all, but they thereby render their own motions unintelligible.[28]

It seems more likely that the 'real world' be taken as the ultimate sum of life-worlds, the total story (which we have not yet learnt to tell), than that it be an unmeaning and strictly indescribable abyss. Uexkuell indeed concluded that astronomy itself was a biological science, concerned with points of light displayed within a human *Umwelt*. That doctrine, even though it was endorsed by Frank Ramsey,[29] does not appeal to me. If Rorty is right to identify his own position as a response to the supposed collapse of Platonism, perhaps it is time that we reconsidered that collapse. Those who would clamber from the cavern of their idiosyncratic dreams, and recognize the power of puppeteers to organize those dreams, may still hope to discover what is really real, not by rejecting what at first appears, but by surpassing it.

Both eliminative materialists and anti-realists deny that possibility, of reaching out to what is other than our dreams and finding it familiar. Both say that the folk-world and its members are only stories told by us, even if anti-realists then go on telling stories, and eliminative materialists pretend to stop. They both seek to spare themselves the possibility of error: anti-realists deny that 'we' could ever be wrong in what 'we' seriously say (for 'being wrong' is only being in the wrong, by our own standards); materialists end by denying that there is any 'we' to be right or wrong. Either way there is no division between what we think and what most truly is. But that risk, of being wrong, is the price we pay for sometimes being right. If it 'makes no sense' to wonder whether people are sentient and pigs are not (because such sentience is either a mirage or — equivalently — a projection of our arbitrary concern), we have evaded a danger, doubtless: but only at the price of surrender. We have lost that sense of Otherness that is the root of love and knowledge. Murdoch's judgment is the better path: 'We take a self-forgetful pleasure in the sheer alien, pointless [?], independent existence of animals, birds, stones and trees .... Good art, not fantasy art, affords us a pure delight in the independent existence of what is excellent.'[30] Art is not all that suffers when we forget that excellence. We have not learnt the right lesson from the private language argument: where there is no chance of error, nothing has been said. So we must hope to mould our thought to what is genuinely Other than our thought. 'To be fully human is to recognize everyone and everything in the universe as both Other and Beloved, and ... this entails granting that the world is authentic and meaningful without demanding proof ... Animals are the only non-human Others who answer us.'[31]

What is outside the text, and much to be desired? Each transformation or escape from our own private story may be welcomed, every realization that the world is wider than our hearts. Why else have we so often imagined fairies, angels, aliens to give a new perspective on a world, our world, grown old? The great discovery we have now almost made is that there are indeed other perspectives on all our human world, that we are the objects of a patient or impatient gaze from animals that share our world and story. It is indeed a difficult task to see through the merely conventional animals with simple moral properties (inquisitive and imitative apes, greedy pigs, proud peacocks, cruel wolves). To evade those traps it is even worth adopting — temporarily — as physicalist a description of the actual behaviour of the animals as we can manage. But it would be as absurd (and perhaps as damaging) to settle for those descriptions as it would be for ardent bachelors to describe the actions of young females 'physicalistically', so as to avoid imputing motives and desires to them that are really only the males' own. To turn aside from the discovery that there are really Others in the world that we can come to know, and to pretend that pigs, dogs, pigeons, people are only pretence, only cuddly toys animated solely by the stories that we (who?) tell with them, is a radical defeat, as dreadful as the other error, which is to pretend to a romantically pessimistic doctrine that we could never ever find out real truths (except that one?). Fortunately for us (and maybe for our immortal souls) neither cats nor infants nor our adult neighbours are so tractable as to support either fancy. Yes, there really is a real world 'out there', and the fact that it is so often not what I would wish is exactly what reveals, and endears, it to me! And yes, that world is the story we shall all have been telling from the very beginning.

## Notes and references

[1] Some of this material has been used, to different effect, in 'The Consciousness of Animals' in H. Robinson & R. Tallis(eds), *The Pursuit of Mind* (Carcanet Press: London 1991), pp.110-28.

[2] R. Bulmer, 'Why the Cassowary is not a Bird' in *Man*, 2.1967, pp.5-25 (discussing the taxonomy preferred by the Karam people of New Guinea). Other Karam folk-taxa include flying birds and bats (*yakt*), dogs, pigs, rats from homesteads and gardens (*kopyak*), frogs and small marsupials and rodents other than *kopyak* (*as*), tadpoles, weevils and snails. I doubt if our own folk-taxonomy is much more rational.

[3] D.A. Dombrowski, *The Philosophy of Vegetarianism* (University of Massachusetts Press: Amherst 1984), p.129, summarizing R. Rorty *Philosophy and the Mirror of Nature* (Princeton University Press: Princeton 1979), p.182-92. Words within square brackets are my own summary of what conventionalism must mean.

[4] Rorty, *op.cit.*, p.190.

[5] I here draw on material prepared for 'Apes and the Idea of Kindred': P. Singer & Paola Cavalieri (eds), *The Great Ape Project* (Fourth Estate: London 1993).

[6] See E. Sober, 'Evolution, Population Thinking and Essentialism' in *Philosophy of Science*, 47, 1980, pp.350-83.

[7] On which see my *Aristotle's Man* (Clarendon Press: Oxford 1975), ch.2.2, and 'Aristotle's Woman', *History of Political Thought* 3 (1982) pp.177-91.

[8] D. Hull, 'Are Species Really Individuals?', *Systematic Zoology*, 25 (1974) pp.178-91.

[9] W.V. Quine, 'Natural Kinds': *Essays in Honour of C.G.Hempel*, (ed.) N. Rescher (D. Reidel: Dordrecht), pp.5-23.

[10] I. Kant, *Gesammelte Schrifften*, vol.2, pp.427-33, cited by Baker, *op.cit.*, p.81.

[11] For further details of modern taxonomic practice, see E. Mayr, *Principles of Systematic Zoology* (McGraw-Hill: New York 1969); C. Jeffrey, *Biological Nomenclature* (Edward Arnold: Cambridge 1973); C.N. Slobodchikoff (ed.), *Concepts of Species* (Dowden, Hutchinson & Ross: Stroudsberg, Pennsylvania 1976).

[12] *Questions sur l'Encyclopédie*, cited by Baker, *op.cit.*, p.20.

[13] All these are genera, with several species usually included in them (as *Pan satyrus paniscus*, the

71

pygmy chimpanzee). Linnaeus identified the chimpanzee instead as *Homo troglodytes*. Nowadays, the level of the taxon is determined by professional judgment having to do with presumed ancestry and relative degree of relatedness.

[14] See 'Slaves and Citizens' in *Philosophy*, 60 (1985) pp.27–46, and 'Animals, Ecosystems and the Liberal Ethic' in *Monist* 70 (1987), pp.114–33.

[15] D. Hull, *Philosophy of Biological Science* (Prentice Hall: Englewood Cliffs, New Jersey 1974), p.79.

[16] R. Kipling, *Collected Verse, 1885/1926* (Hodder & Stoughton: London 1927), p.547.

[17] The two condemnations, to be fair, are not entirely incompatible. Some would say that what was relevant was simply pain. Sacred cows, because nobody kills them, suffer protracted deaths; Korean dogs perish by slow strangulation.

[18] See V. Hearne, *Adam's Task: Calling Animals by Name* (Alfred A. Knopf: New York 1986).

[19] O. O'Donovan, *Begotten or Made* (Clarendon Press: Oxford 1987).

[20] G.K. Chesterton, *The Poet and the Lunatics* (Darwen Finlayson Ltd: London 1962; first published 1929), p.70.

[21] Chesterton *op.cit.*, p.58.

[22] C.S. Lewis, *The Abolition of Man* (Bles: London 1946, 2nd ed.), p.12.

[23] See my 'On Wishing there were Unicorns': *Proceedings of the Aristotelian Society*, 1989-90.

[24] A.MacIntyre, *Against Virtue* (Duckworth: London 1981), p.201.

[25] J. von Uexkuell, *Theoretical Biology*, tr. D.L. Mackinnon (Kegan Paul: London 1926). See also his 'A Stroll Through the Worlds of Animals and Men': C.H. Schiller, (ed.), *Instinctive Behaviour* (International University Press: New York 1957), pp.5-80.

[26] D. Bloor, *Wittgenstein: A Social Theory of Knowledge* (Macmillan: London 1983), pp.174-6, points out that Wittgenstein's observation is founded in a 'quite proper sense of the biological basis of social life'.

[27] W. James, *The Principles of Psychology* (London: Macmillan 1890), p.288.

[28] See the article previously cited, in Bekoff & Jamieson.

[29] Uexkuell, *Theoretical Biology*, *op.cit.* pp.35ff; F.P. Ramsey, *Foundations of Mathematics* (Kegan Paul: London 1931), p.291: 'I don't really believe in astronomy, except as a complicated description of part of the course of human and possibly animal sensation.'

[30] I. Murdoch, *The Sovereignty of Good* (Routledge & Kegan Paul: London 1970), p.85.

[31] Hearne, *op.cit.*, p.264.

# The thinking animal

## Marthe Kiley-Worthington

*Scholarly research on animals and our attitudes and understanding of them, must, by definition, be inter-disciplinary. To progress in this field towards post-modern Zoology thus, it is necessary to combine knowl-edge from many disciplines and to understand that science must be used in its original meaning: the understanding and acquiring of rationality and knowledge.*

(Oxford English Dictionary 1988)

Such a study has the potential of developing into a most exciting and challeng-ing branch of 'post-quantum' endeavour. However, to achieve this, it will have to be careful that it does not show an excessive respect for only empirical results and ignore others, be morally suspect (e.g. by causing animals to suffer while making pointless empirical measures) and intellectually dull. If science is to be the pursuit of knowledge rather than reflect cultural attitudes and employment for technicians, it will be necessary to consider many questions that have not been asked by scientists.

In the first place, I suggest, there are some widely held and unjustified assumptions concerning animal minds and consequently our attitudes to them which, until recently at least, have escaped serious investigation. The first one of these is that since emotional responses of animals have been difficult to define and quantify they have often been ignored or have received little attention from ethologists and others interested in animal behaviour, even though we may have considerable a priori knowledge of them. We know that our dog experiences misery and joy, and we know when, but as scientists and investigators we have been guilty of ignoring this 'folk psychology' or 'common sense knowledge'. This to a large extent has been a matter of convenience. If we are seriously aware of the similarities in emotional responses to those of humans and their strength and complexity in different animals (even the animals we keep and eat such as pigs, chickens and cattle, as well as those we use in many other ways), then it can become a matter of embarrassment to justify the treatment we often give them. Our lives would be much less easy in many ways if we were to consider the animals and humans emotional needs as similar!

Secondly, there has been in zoology as in other sciences, and still is in some areas, a confirmed belief in the possibility of 'objective assessments'. Yet at the same time, even the questions asked will reflect the subjective interests of the observer, and in turn the cultural beliefs. As far as animal behaviour is concerned, there are many examples of this. We will take one; the idea of 'dominance hierarchies'. The theory here is that individuals in animal societies will inevitably be in conflict with one another, and the only way they are able to live together is if every individual 'knows his place', that is has a particular 'dominance status' in the inevitable hierarchy. Thus researchers go out into the field with preconceived notions of the society they try to understand: to establish the 'dominance hierarchy' and even in some cases on returning to the laboratory to convert their measures into mathematical exercises: angles, logarythmic scales and so on.

Few if any researchers question whether or not the assumption of competition and conflict and thus the existence of 'dominance hierarchies' is reasonable or correct. After all it would be quite possible, and indeed might be even more likely, that individuals stay together in groups because they co-operate and like each other. postmodern zoology is however beginning to examine such ideas (e.g. Mae-Wan Ho 1989, Capra 1982).

Thirdly, and most importantly, is the rarely questioned assumption that human beings have greater abilities and more important interests than other species.

This results in other sentient beings being assigned little or no importance except to serve human ends. In other words they are assumed to be behaviourally inferior, and as a result of no moral worth and have no interests (e.g.

Frey 1983, Paton 1984). Others who are concerned with the morality of our treatment of animals will argue for their moral worth, despite their behavioural inferiority (e.g. Rachels 1989, Midgeley 1983, Clarke 1977). The question we will address here, and perhaps the central concern that will form our postmodern zoological understanding is *are we justified in assuming other mammals at least, are behaviourally inferior to humans?*

One way of doing this is to look at the way humans relate to animals. The most common attitude, although rarely admitted is that of Exploitation/Parasitism.

Through history animals' association with and husbandry by humans has been characterized by their exploitation: human parasitism at the animals' expense; or at best, a relationship of master and slave. Examples in all types of animal husbandry systems are too numerous to need quoting, from companion animals, to farms, zoos, circuses, working, sport and laboratory animals. This exploitive position taken by humans has been retained often unquestioningly because of convenience, and has resulted in many religious and cultural justifications (e.g. Frey 1983, Paton 1984, Whale 1965, Humphrey 1984, Serpell 1986), although which is cause and which effect has not been clearly established.

A commonly held position is that as a result of not having 'language' (the definition of which remains disputable) animals are not of moral concern and thus can be exploited. The first three-quarters of the twentieth century was characterized among most ethologists and psychologists by the belief that it was scientifically unacceptable to state that other mammals could feel, think, suffer emotions or even pain. If they did suffer pain, it was a different pain from that of humans (Iggo 1987, Frey 1984)

It would seem that the time has now arrived, to take cognizance of 'common-sense knowledge' (Rollin 1989): we know dogs (who like us are mammals) feel emotions, just like they eat, drink and reproduce. To be good scientists we do *not* have to assume they do not until it is empirically tested. The bonus is on those who consider that humans and other species are qualitatively different in their abilities to experience emotions and thus not the result of similar evolutionary rules to prove that they are different. It is irrational and bad science to ignore 'common-sense knowledge', the problem is how to define it.

The conventionally believed distinctions of mind between humans and other animals have been maintained in Western societies for several thousands of years: humans are the only ones who have souls, who think, are self-aware, have language, who can work out and use rational solutions, have an idea of the future and the past, have different cultures, have an aesthetic sense and who behave altruistically. These beliefs are now being carefully examined and tested, and as a result of work by many including ethologists, psychologists, cognitive scientists and philosophers we are realizing that such statements are no longer acceptable at face value.

There is still controversy on these issues, but in the last decade, it is now accepted that animals feel emotions, can and often do act rationally, show forethought, have or can learn relatively complex 'languages' (Savage-Rumbaugh 1977, Premack 1986, Gardner & Gardner 1975), can learn concepts (Herrenstein *et al.* 1984, Watanabe *et al.* in press, Dashevsky 1991). Indeed it is evident that each species, including humans, has particular receptor, communication and cognitive abilities: unique minds. To assume, for example, that humans' receptors are necessarily more elaborate, sensitive and complex (superior) is demonstrably false. Can we then be *so sure* that all aspects of cognition and mental experience of humans are necessarily more elaborate, sensitive and complex (superior), than other species, or would this not be foolhardy?

So the question of the individual animals cognitive abilities is suddenly one that cannot be dismissed in terms of 'instinct': no thought, no desires, no intentions, no beliefs. Fodor (1977) talked about the 'language of thought': 'propositional attitude of tokens (symbols) which have intentional content'. Churchland (1984) goes one step further: 'different degrees of self-consciousness are possible depending on the practice and experience of the animals. It therefore has a large learnt component ... Therefore it exists in some degree in every cognitively advanced creature.' This is of course something all good animal trainers have taken for granted for a very long time. Xenophon (300 BC) states that when educating horses their *understanding* must be developed first. Without assuming that animals as well as children, think, feel, and learn, trainers and teachers would not be able to train animals or children. Postmodern zoology must progress in this knowledge, not ignore it.

Animals mentally perform very complex tasks, for example whether or not the chimpanzees' mastery of American Sign Language, and use of symbols with computers is called 'language' (e.g. Premack 1986, Savage-Rumbaugh 1977) matters little; the point is they can do it. Pigeons have been taught to make relatively complex decisions which we had no idea they might be able to do: understand the concept of a tree (when is a tree a tree, it may have leaves or not, be of a variety of shapes and colours), and the difference between a tree and a bush. Artificial concepts have also been used, such as a hat: hats come in all sorts of sizes, colours and shapes and they are, one would imagine, somewhat irrelevant to most pigeons or apes, yet they can learn the functional discrimination of the concept of Hat. I wonder if we could do the same for their worms and bugs? Pigeons also manage to remember up to 300 slides, make correct judgments concerning views taken from a range of angles, and so on.

A parrot has been taught to speak English and say what he means. Shown a tray of objects of different shapes, colours and made of different materials, Alex, the African grey parrot, will pick out and describe in English 'the blue triangle made of wood' for example (Pepperburg 1991). Dolphins have learnt the concept of novelty. They also empathize towards stranded humans in the sea, and help them; one old legend of the sea is turning out to be sometimes true. Dolphins learn such behaviour quickly and apparently are willingly to take part in helping mentally and physically handicapped humans to swim with them.

The animals most of us have most to do with, such as dogs and cats, horse, cattle, pigs and chickens, have yet to be tested in a similiar fashion but there is little reason to assume that they are not able to do these mental analyses if pigeons and rats can.

The mind/brain relationship and the interrelationship of intellectual and emotional abilities remain controversial, but recent animal cognitive research has given them a new lease of life and made us re-examine many of our preconceived convenience and cultural beliefs about the cognitive abilities of other animals. There is little doubt now that mammals and birds at least, and probably many other animals too, have desires and beliefs, show intentionality, learning, rationality, sophisticated communication and social behaviours. In other words they are in many ways mentally similar rather than totally different from humans. It is true to say that humans are probably the most manipulative mammal, and have a particular linear way of problem-solving; but does this mean this is the best and *only* way of achieving any cognitive goal? Clearly not, there may well be a great many more ways of thinking, problem solving, experiencing the world and decision-making that we are not yet aware of, and it may be that we do not have to look to outer space, science fiction, the pacific ocean or tropical rain forests for experts in this. They are here on the mat in front of the fire: Fido, Bruno and the rest.

What sort of mumbo-jumbo then are statements such as: 'no chronic distress can be suffered by laboratory animals because they have no preconceived notions.' This may be true for animals or uninformed humans the first time they have an unpleasant experience in a particular set of conditions, however, all mammals and birds anyway certainly have 'preconceived notions of unpleasant experiences' after one experience. This has been horrifyingly demonstrated by too many psychologists. Seligman (1970), for example, turned dogs psychotic by continuing to shock them when they expected it and could not escape ... Is this science?

In order, therefore, to avoid drawing biased conclusions that are the result *predominantly of our cultural education and prejudices* rather than truly scientific, we must conclude: (1) Warm-blooded species, including humans, other mammals and birds, have brain, behavioural, physiological and morphological similarities. Superimposed on these overall general similarities are species, breed, sex and individual differences. (2) That each species is unique, each has a different social organization, lifestyle and habitat choice, repertoire of behaviour, mental experiences and ecological requirements. (3) That individuals have (a) genetic and (b) lifetime experience differences which affect their behaviour and assessment of the world.

The rational development of these considerations does not lead to animal exploitation and parasitism; rather, the reverse. Indeed, perhaps we could advance knowledge (science) if we were to look at other species not as emotionally and cognitively less able 'inferior beings', but rather as an enlightened anthropologist might look at another human culture: from which we have the option of learning more about both the other species and ourselves, and the limits to our interpretation of the world as a result of 'la condition humaine'.

A common attitude to animals that has arisen in the last decade or two since debates on animal welfare and conservation hit the headlines is that of Animal Apartheid. It is often argued correctly and rationally perhaps (e.g. Singer 1976, Regan 1983, Saponzis 1987) that we should show equal consideration to other sentient beings. However, the consequence of this position has to date been that it has been assumed that different species must live exclusively in their own societies in the 'Wild Yonder': Nature Reserves, for example, into which humans only venture occasionally and then as very considerate guests. This position assumes that contact with humans must necessarily be bad for the animal and thus separation is best: Animal Apartheid.

There are two main objections to Animal Apartheid. The first assumption underlying it is that it is not possible for humans to use animals, or be used by them symbiotically, therefore they must be kept apart. With expotentially growing human populations, this is becoming a greater problem everywhere. Inevitably such 'animalistans' will become more and more restricted, and consequently fewer species will survive (e.g. Erlich and Erlich, 1982). The second more serious assumption is that interaction between humans and other animals must inevitably result in human exploitation and parasitism. The people who hold this view include great many of those involved with conservation and environmental matters. Yet we do recognize dependency relationships within the biosphere. Humans have a particularly manipulative role at the moment, but does this require that they must be kept apart from the rest of the biosphere for each party to benefit?

Being brought up in a homocentric, even 'homoexclusive', environment is less likely to result in an understanding of the biosphere and its central importance to human beings. It is also obvious that human beings might learn more and different things about the world and how to interpret it from close harmonious interaction and mutual education with other life forms and sentient mammals in particular. Four years living with and studying cattle allowed me to collect much empirical data on many aspects of their behaviour, but as a result of the protracted period with them I inevitably absorbed other information on the way in which they interpreted the world. For example, their concept of time, and their ability to accept the world as it is rather than as something to change (the way in which some species often consider it: humans, some primates and horses). Gathering information on inter-species cognitive differences and similarities and their 'realities' is perhaps the stuff postmodern zoology will be made of.

There is another possible attitude to, and relationship with, animals. This is 'Animal/Human Symbiosis'. If we are rational scientists knowing what we do now about animal minds, we must entertain the possibility that both the animals and the humans might benefit from some interaction. Thus they might be able to come to some *symbiotic* relationship rather than one characterized by human exploitation and parasitism, or Animal Apartheid.

This is conventionally considered possible with domestic and companion species, but not for wild animals who are given different status. The reason why this is so, it is said, is that domestic species are somehow fundamentally flawed because their breeding has been interfered with by humans, and their behaviour, it is argued, has been genetically changed. The truth of it is that fundamental behaviours have not changed dramatically or even hardly at all (Kiley-Worthington 1977, Price 1984).

Lifetime experiences greatly affect behaviour of all species. It may be that we have exerted some selection pressure on domestic species that facilitiates their abilities to co-operate with humans, to learn human communication systems, and increased adaptability. It is simply false to state, for example: 'wild animals have an instinctive fear of man, whatever their age and experience.' Unless, as Hearne (1986) observes: 'instinct is an odd term for a large collection of abilities, including keen observation and analysis!' No, assumed generalizations concerning the needs and desires of all wild animals as *qualitatively different* from all domestic animals as a result of *inherited* differences in behaviour is false.

Similarly the widely held belief in genetic differences in behaviour between domestic breeds may not be straightforward. Take dogs, for example. There has been artificial selection for a great variety in size, colour and shape of domestic dogs, but to what extent does behaviour vary between breeds *genetically*? Do we have evidence uncontaminated by lifetime experiences of, say littermate puppies? The answer is no, we do not have *genetic* evidence that

Rotweilers and Pit Bull terriers are more aggressive than other breeds. To test this, we will need different littermates of different breeds being raised in different ways by different humans who have no preconceived notions on how the different dogs should behave. We certainly do have evidence that different humans raise differently and this results in different behaviour from the dogs.

Indeed why should we be so convinced that different breeds of domestic dogs have very different behavioural characteristics when we do not, generally, belief this for different breeds or races of humans (there are some who do: they are usually called fascists). Why should dogs be controlled by different genetic rules to humans?

In the light of recent ethological and cognitive knowledge of animals, the old-style belief which held that other sentient animals can be discriminated against and used only for human ends without further discussion, even though the animals suffer as a result, is irrational and non-scientific (e.g. Paton 1984, Iggo 1988).

If there is any substance to the argument that it might be possible to live symbiotically with other animals, wild or domestic, how is it to be achieved? When are we to say that it has been achieved and that the animal and the human have both benefited? In the first place we must know when we have *not* achieved this from the animals' point of view. Can we have basic guidelines which apply animal husbandry systems to ensure that they do not suffer (at least for prolonged periods), and more, that they feel pleasure and joy when they are kept in certain husbandry systems by human beings? If so, what are these?

Secondly, how can we work towards greater interaction and communication, and thus greater mutual understanding and an enrichment of life between humans and other species to approach symbiotic living?

We already have sufficient knowledge to be able to assess distress and suffering in animals and in cases of doubt we should give the animal the benefit (Duncan 1987, Fox and Mickley 1987, Kiley-Worthington 1983, 1990; among others). Since we, humans, are also mammals where there is controversy or a lack of evidence it seems better science to identify and use mammalian similarities — conditional anthropomorphism in the first instance, when making judgments concerning animal suffering. Thereafter, where we have sufficient knowledge, it may be possible to make pertinent distinctions between species. Guidelines for assessing distress and suffering are possible (Fig. 1).

Physical suffering (pain, wounds, starvation, neglect resulting in malnutrition, disease, etc.) are generally agreed criteria which are routinely used. In addition, incidence of common physical occupational diseases (for example lameness in jumping horses), and the frequent or persistent need for drugs (e.g. the routine feeding of antibiotics) or surgery (the cutting off of the beaks of chickens, tails of pigs to prevent cannibalism, or of dogs for cosmetic reasons, castrating of males, etc.) in order to enable the maintenance of the animals in a particular system, or as a result thereof, should also be assessed. Psychological ill health: prolonged distress in different species is more difficult to assess but there is a substantial amount of knowledge now from both humans and other animals to allow us to err on the side of the subject.

These guidelines may rarely be fulfilled in almost every animal keeping enterprise, but they are, perhaps, what we should work towards if symbiotic relationships between species are to be achieved. One of these criteria that is used almost universally for humans is that of Behavioural Restriction. If humans and other mammals are cognitively more similar than we thought, why is it that without a qualm we take infants away from their mothers, shut isolated animals up, and or castrate many individuals? We would not consider there was justification for doing this to humans, so why do we assume that is not cruel and causing distress to other mammals ?

Evolutionary biologists (which encompasses almost every biologist to date) consider that there are functional reasons for all normal behaviours which have been selected by natural selection. Thus it would seem rational to consider that the animal has a physical and/or psychological 'need' to perform them (Thorpe 1962). Consequently, to optimize her welfare, an individual animal should be in an environment where she has the opportunity to *'perform all the behaviours in her repertoire which do not cause prolonged suffering to others'*. We have *a priori* knowledge that the animals 'need' to perform all the behaviours in their species repertoires because they have evolved, and the animals do them: ducks swim, copulate and associate together on water, so they *need* water not just to drink. Hens flap their wings, dust bathe, move about, and select places to lay their eggs. The priority the animal (or human) gives to these behaviours varies with time and contextual stimuli as well as with different species and individuals. For example, from moment to moment the hierarchy of your, mine and Jumbo's behavioural needs may change: if hungry we will eat, if sleepy sleep, if social interact, if we have the chance. How can we make judgments concerning what is a 'real' need and what is a 'luxury'? Analogies with *a priori* knowledge on human behavioural needs is rational. We cannot have indications of which behaviours are more or less important to the animal overall. It is likely for example that sex, maternity or appropriate social contact may be *more* important for any animal than food or water at certain times — as it is for humans. Statements such as: 'copulation among cats is an *instinctive* behaviour touched off solely through the physiological changes in females during periods of oestrus ... There is no logical reason not to spay your female — none.' Sautter and Glover, 1978 assume that:

(a) Cats do not have communication or foresight which will therefore make them aware of missing out on sex. We may have no evidence that they *have* but we cannot therefore rationally assume they *have not*.

(b) If experiences are never had, they will make no difference to the quality of life — 'out of sight is out of mind'. If this were the case it would then presumably be an easy solution to human population problems — castrate or hysterectomize all young human infants that are not required for breeding and keep them isolated from the rest of the population. This is not usually considered an acceptable solution as we consider these individuals will miss out on many experiences, even though they may not have been told about them.

Humans and other animals are not *tabula rasa*; they inherit certain behavioural tendencies to do various things and behave in certain ways. Certainly these are modified by present and past experiences but it is irrational to prevent them from doing them without very considerable thought ... if we are interested in their well-being.

What is possible and necessary now, if we are really interested in animal/human symbiosis, is to make assessments of the relative behavioural options in different environments: a 'behavioural restriction quotient'. This is not at present done; traditional behavioural restrictions are rarely questioned. For example, why is it assumed that it is acceptable to separate dairy cows from their calves within forty hours of the calf's birth, and never to allow cows to have sex? Reduction of behavioural restriction would be possible within almost every category without abandoning that type of husbandry (Fig. 2).

If the animal is able to experience distress and suffering, then she should also be able to experience joy and pleasure. I have made a start with trying to assess joy or pleasure in different species in circuses and zoos (Fig. 3).

Can we with existing knowledge set up the Optimal Environment for our animals? We can come closer to it by considering many ethological criteria including the individual's 'telos' (Aristotle 350 BC): the elephantness of the elephant; the horseness of the horse. These will be:

(i) physical needs;

(ii) emotional/social needs;

(iii) cognitive needs, including 'intellectual' and possible educational needs;

(iv) individual needs that are the result of past experiences.

It is thus possible to draw up guidelines for symbiotic living with animals, or even for the types of environments they could live in without us: Ethologically Sound Environments for animals (Fig. 4).

There is little knowledge on species differences in learning, cognition, perception and brain anatomy except at a very gross level.

Consider for a moment Fig. 5. Here the brains of four higher mammals are shown, now ignoring the importance of the body to brain size or weight ratio (which is controversial anyway), look, if you will, at the fore brain: the cerebral cortex of these four. The dog, widely believed to be man's best friend and considered the most intelligent of our domestic animals, has a relatively small surface area to the fore brain (the part that is considered to be mainly concerned with cognition, thinking, rational thought and so on). By contrast the fore brain of the cow and the horse is extremely well developed and highly convoluted. In fact there are more convolutions in the fore brain of the horse than that of the human brain ... whatever that means. What this does seem to indicate is that the fore brain of cattle is more developed than one would consider it needed to be if these animals only live the sort of life we think they do, a bit of socializing, sex, eating, ruminating and wandering around ... they would not need such a big fore brain ... after all, the socially hunting carnivor, the dog, does not need much (Fig. 5.1.). So what is it for? And what is going on in there?

Because — as far as we know — animals, even these higher mammals, do not have the possibility of learning intellectual skills from previous generations by talking, reading and so on, and because so far we have assumed that cattle and horses are of very limited mental ability and we have rarely, if ever, spent even one tenth of the time teaching them that we spent on our own children, the question must be asked: what would happen if we put the same time and motivation into raising and educating a horse or a cow as we do for our own infants? Would we be able to learn from them a little more about their world, their thinking, their intellectual needs and abilities — their cognition?

Such are the questions that will be part of postmodern zoology, leading, perhaps, to Chiron's World: animal/human symbiosis.

We are, after all, all citizens of the world.


## Bibliography

Aristotle, *Basic Works,* (ed.) R. McKeon (Random House: New York n.d.).

F. Capra, *The Turning-Point. Science Society and the Rising Culture* (Flamingo Press: London 1982).

P.M. Churchland, *Matter and Consciousness* (MIT: Mass. 1988).

S. Clark, *The Moral Status of Animals* (Clarendon Press: Oxford 1977).

B.A. Daskevsky, *Perceptive Learning and Cognitive Spatial Capabilities* (Int. Ethol Congress: Kyoto, Abst. 1991), p.151.

R. Erlich & A. Erlich, *Extinction* (Victor Gollanz: London 1982).

J. Fodor, *Psychological Explanation* (Random House: New York 1968).

M.W. Fox & L.D. Mickley (eds), *Advances in Animal Welfare Science, 1986-7* (Nijhoff: Boston 1987).

R.G. Frey, *Rights, Killing and Suffering* (Blackwell: Oxford n.d.).

B.T. Gardner & R.A. Gardner, 'Evidence for Sentence Constituents in the Early Utterances of Child and Chimpanzee' in *J. Expt. Psych. General* (1975), 104, pp.244-67.

V. Hearne, *Adam's Task. Calling Animals by Name* (Heinemann: London 1986.)

R. Hernstein, D. Loveland & P. Cable, *1976,* 'Natural Concepts in Pigeons' in *J. Expt. Psychol. Animal Behaviour, Proceedings* 2 (1976), pp. 285-302.

Iggo, *Pain in Animals* (University Federation of Animal Welfare: Potters Bar 1984).

M. Kiley, *Behavioural Problems of Farm Animals* (Oriel: Stockton 1977).

M. Kiley-Worthington, 'Ecological, Ethological and Ethically Sound Environments for Animals' in *Journal of Agricultural Ethics 2* (1989), pp.323-47.

M. Kiley-Worthington, *Animals in Circuses and Zoos. Chiron's world*?? (Little Eco-Farm publ.: Basildon 1990).

*Oxford English Dictionary* (OUP: Oxford 1988).

Mao-Wan Ho., *Theoretical Biology: Epigenetic and Evolutionary Order from Complex Systems,* (eds) B. Goodwin & P. Saunders (Edinburgh University Press: Edinburgh 1989).

M. Midgeley, *Beast and Man. The Roots of Human Nature* (Methuen: London 1979).

W. Paton, *Man and Mouse* (OUP: Oxford 1984).

J.M. Pearse, *An Introduction to Animal Cognition* (Lawrence Erlbaum: London 1987).

I.M. Pepperburg, 'A Communication Approach to Animal Cognition, a Study of Conceptual Abilities of an African Grey Parrot' in *Cognitive Ethology,* (ed.) C.A. Ristau (1991), 'pp.153-77.

B.O. Price, 'Behavioural Aspects of Animal Domestication' in *Quarterly Review of Biology* 59 (1984), pp.1-32.

D. Premack, *Gavagai* (MIT: Mass. 1986).

J. Rachels, *Created from Animals: The Moral Implications of Darwinism* (OUP: Oxford 1991).

T. Regan, 'Animal Rights Human Wrongs' in *Ethics and Animals,* (eds): H.B. Miller & W.H. Williams (1983), pp.19-44.

B.E. Rollin, *The Unheaded Cry: Animal Consciousness, Animal Pain and Science* (OUP: Oxford 1989).

S.F. Saponsis, *Morals, Reason and Animals* (Temple University Press: Philadelphia 1987).

E.S. Savage-Rumbaugh & D.M. Rumbaugh, 'Symbolism, Language and Chimpanzees: Theoretical Reevaluation Based on Initial Language Acquisition Processes in Four Young Pan Troglodytes' in *Brain & Language* 6 (1978), pp.265-300.

F.J. Sautter & J.A. Glover, *Behaviour, Development and Training of the Cat: A Primer of Feline Psychology* (Arco: New York 1978).

M.E.P. Seligman, *On the Generality of the Laws of Learning, Psychological Review,* 1970, pp.406-18.

J.A. Serpell, *In the Company of Animals* (Blackwell: Oxford 1986).

P. Singer, *Animal Liberation* (Jonathan Cape: London 1976).

S. Watanabe, S.E.G. Lea & W.H. Dittrich, 'What Can we Learn from Experiments on Pigeon Concept Discrimination?' in *Avian Vision & Cognition,* (eds) H.J. Biscopf & H.D. Zeigler (MIT: Mass. forthcoming).

J.S. Whale, *Christian Doctrine* (Routledge & Keagan, Paul: London 1965).

Xenophon, *The Art of Horsmanship* [300 BC], trans. M. Morgan (J.A. Allen: London 1962) .

## FIGURE 1: INDICATORS OF DISTRESS IN ANIMALS

* Evidence of physical ill-health (including poor nutrition, wounds, etc.)
* Evidence of frequent occupational diseases
* Need for the use of drugs and/or surgery to maintain the system of husbandry
* Behavioural changes :
  (a) performance of abnormal behaviours (that are not normally in the animals' repertoire, and which appear to be of little benefit to the animal: e.g. running at bars, pacing)
  (b) stereotypies, i.e. the performance of repeated behaviour fixed in all details and apparently purposeless (e.g. crib-biting, wind-sucking, weaving, head twisting)
  (c) substantial increase of inter- or intra-specific aggression compared to the wild or feral state
  (d) large differences in time budgets from the wild or feral animal
  (e) substantial increases in behaviour related to frustration or conflict (e.g. often behaviour relating to locomotion and/or cutaneous stimulation)
  (f) substantial ontogenic behavioural changes (animals performing behaviour characteristics of a very different time in their development, e.g. calves of sixteen weeks walking as if they were a day or so old)
* Behavioural restrictions: this is the inability to perform all the behaviour in the animals' natural repertoire which does not cause severe or prolonged suffering to others.

## FIGURE 2 : THE BEHAVIOUR RESTRICTION OF DAIRY CATTLE IN DIFFERENT HUSBANDRY SYSTEMS

| | Feral or wild | Pasture & shelter with bull | Straw yard & run, calves & bull | Straw yard no run/calves/bull | Cubicle, no run calves/bull | Yoked or tied no calves/bull |
|---|---|---|---|---|---|---|
| Never unenclosed | | * | * | * | ** | *** |
| Never unenclosed | | * | * | * | ** | ** |
| Self-comfort | | | | * | *** | *** |
| Choose social partner | | | | ** | * | *** |
| Mixed sex & age | | * | * | ** | ** | ** |
| Sexual behaviour | | ** | | ** | ** | ** |
| Maternal behaviour | | ** | | ** | ** | *** |
| Dull environment | | * | | * | ** | *** |
| All gaits possible | | | | * | ** | *** |
| TOTAL | | 8 | 3 | 13 | 19 | 23 |
| Always finder shelter | * | | | | | |
| Water | * | | | | | |
| Food | * | * | | * | * | * |
| TOTAL | 3 | 9 | 3 | 14 | 20 | 24 |

Even within these different environments the amount of behavioural restriction varies considerably with the individual design. This is to show that under human jurisdiction, life for cows is usually very behaviourally restricted (column 2, 4, 5, 6) but it *can* with thought, be at least behaviourally unrestricted as it is for wild or feral cattle (column 3). Such husbandry systems can also be economic. We have been running and living on the proceeds of one for eighteen years!

## FIGURE 3 : BEHAVIOURS WHICH ARE POSSIBLE INDEXES OF PLEASURE

| Elephants | Bovids | Camelids | Equids | Canids | Felids |
|---|---|---|---|---|---|
| Blow | Elevated paces | Elevated paces | Elevated paces | Elevated paces | Elevated stand |
| Bang trunk | Leap | Leap | Leap | Leap around | Leap around |
| Cough | Chase | Chase | Chase | Chase | Chase/roll |
| Mutual groom | Mutual groom | Mutual groom | Mutual groom | Mutual groom | Mutual groom |
| Play | Play | Play | Play | Play | Play |
| Rumble | Tail elevate | Tail elevate | Tail elevate | Tail elevate/wave | Tail wave |
| Stretch | Stretch | Stretch | Stretch | Stretch | Stretch |
| Ear flap | Ear prick | Ear prick | Ear prick | Ear prick | Ear prick |
| Leap | Rush | Rush | Rush | Rush around | Rush jump |
| Touch other | Touch other | Touch other | Touch other | Touch rub other | Touch rub other |
| Squeal | Purr | Speak | Squeal/nicker | Lick other | Lick self & other |
| Trumpet | Gambol | Gambol | Buck/kick | Gambol | Gambol |
| Roll | Doze | Doze | Doze | Doze | Doze |

Adapted from Kiley-Worthington, 1990

89

FIGURE 4 : CRITERIA FOR ETHOLOGICALLY SOUND ENVIRONMENTS FOR ANIMALS

(1) MINIMAL BEHAVIOURAL RESTRICTION
The animal should be able to perform all the behaviour in his repertoire which does not cause prolonged or acute suffering to others.

(2) APPROPRIATE PHYSICAL ENVIRONMENT
The animal should be kept in environments physically ressembling that which he evolved to live in (for example, forest-dwelling animals in a forest, or simulated forest; plains-dwelling in an open environment; with or without light as appropriate).

(3) APPROPRIATE SOCIAL GROUPING
The animal should be allowed to associate in the type of social group, size and structure he has evolved to live in.

(4) A CONSIDERATION OF THE ANIMAL AS REPRESENTATIVE OF A SPECIES
The animal's 'telos': species uniqueness must be catered for by, for example, considering :
(i)   The neuro-physiology of his receptors and brain physiology.
(ii)  The neuro-anatomy of his brain.
(iii) His cognitive and intellectual abilites.
(iv)  A detailed understanding of his specific communication system.

(5) A CONSIDERATION OF THE ANIMAL AS AN INDIVIDUAL
The animal must be considered not only as a representative of a species, but also as an individual with a consideration of his past experience and its effects.

(6) PLEASURE
There must be some understanding of how he displays 'pleasure', and its display assured.

(7) NOT DISTRESS
There must be no evidence of distress.

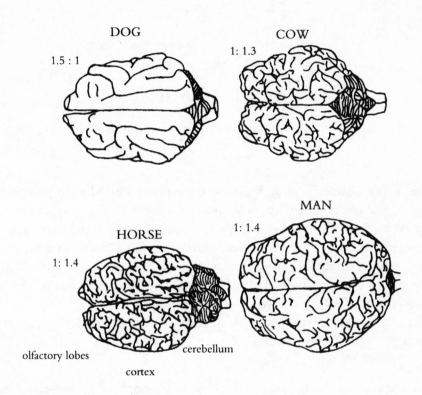

Differences in the gross anatomy of the brains may or may not tell us something about the mental abilities of the species. The size of the fore brain (the convoluted part) is often considered to indicate 'intelligence' and ability for rational thought. If this is the case, then the dog's abilities in this direction are much less than the horse's! Indeed, the horse's cerebrum has more convolutions and therefore greater surface area than that of humans, and the cow's is not much different. What all this means we don't really know, but perhaps we should not underestimate animals' cognitive abilities.

# Animals à la carte

## *Ulrich Melle*

'No Pigs in Press Street': An agribusinessman's plan to build a pig farm on the outskirts of the town met with vehement protest from the neighbours. The continuous noise of the nearby motorway notwithstanding, the town has preserved its rural nature — by Flemish standards. An industrial farm with tens of thousands of pigs would be the equivalent of a waste dump or a chemical plant. Nobody wishes to have this in their back yard. The Tchernobyl of the Flemish pig industry is still fresh in everyone's mind.

    The following quotations are taken from newspaper articles that appeared from January to July 1990 (my italics):

Weelde 18/1
'In a breeding and feeding farm in Weelde (a borough of Gravels), a case of swine fever was discovered. 297 pigs had to be *slaughtered* and *destroyed*. (...) As yet no new cases have been discovered ...'.

Wingene 23/2
'Two new sources of swine fever were discovered in Wingene. (...) It is not yet known how many pigs will have to be *slaughtered*, but it will be well over 1000. (...) In Wingene 200,000 pigs are raised. In the Tielt-Wingene-Ruiselede area, the figure totals 800,000.'

Wingene 11/3
'Another solution that has been suggested is the complete *extermination* of the

165,000 pigs still within the Wingene quarantine zone.'

Brussels 18/3
'The current swine fever epidemic in West Flanders has up to now caused the loss of 40,000 pigs. For an average price of 5000 Belgian Francs, this means a financial loss of 200 million francs [approx. £4,000,000] for the pig farmers.'

Wingene 8/4
'The new measures taken by the Ministery of Agriculture *to slaughter* 150,000 animals in the quarantine zone ...'

Brussels 18/5
'Judging by the situation on 17th Ma , 274,180 pigs will have to be *slaughtered*. ... that the *destructive capacity* is limited ... the preventative *extermination* of all pigs within the 1 km zone ...'

Brussels 12/7
'On 11th July, 89 nidi of swine fever had been discovered. A total of 309,167 pigs have been *slaughtered*. An additional 474,066 pigs and piglets were *destroyed* as a result of the buy-up scheme.

Belgium is home to 21,000 pig farms. Normally they *process* 10 million years pigs per year. The industry represents a productive value of 60 billion francs (in 1989).'

The swine-fever that struck Belgium from January to September 1990 finally cost the lives of over one million pigs — slaughtered, destroyed, exterminated, finished off, or simply having perished in overcrowded sties, suffocating in their own excrement, dying from hunger and thirst or bitten to death by desperate fellow sufferers. Television showed images of heaps of entangled corpses of dead pigs. The industrial farmers were blamed for the epidemic. On the other side stood the family farms; they, it seemed, were the real victims — not the pigs. On the one hand, the large anonymous agro-industrial capital, on the other hand the small and undisturbed world of the traditional breeder on his closed farm, with his year-long devotion to the breeding of sows, with his love for the business and for his animals.

Everyone interested in breeding pigs or in dealing with them, should feel joy and love for it, and has to know the basic characteristics and behaviour of pigs. The following description does not claim to be complete: pigs have superior hearing and scent, a good sense of taste and relatively good power of vision; they have a good memory and learning capacity. They like: companionship and living in a community, to eat with other pigs, regular feeding times, enough to drink, enough sleep and rest, sties that are dry, not too hot or too cold and not too small, preferably straw-littered; they enjoy wallowing in the mud when temperatures are up, grubbing up the ground; they like the unpleasant smell of rotting substances, and expect gentle treatment by the people who care for them ...[1]

In short, pigs are living beings, social mammals with instincts, abilities and needs typical for their species, who are able to individually develop their possibilities as a pig and to enjoy the completion of a life that was tailor-made for them. Grunting pigs on a field with a mud pond and trees appeal to our feelings: they present an image of the comfort and happiness of a living creature. A pig in a modern family farm may be better off than its counterpart in an industrial pig farm, from a hyoistic point of view both are part of the pig Gulag. But then who would be willing to identify even for one single moment with a pig and its needs?

It should be admitted that modern stock-breeding methods pose enormous problems to ontology and ethics. It is unclear under which category of being a modern store pig should be classified. It is a post-natural being, a kind of living machine with only one purpose: maximum growth. The problems are even bigger with the transgenous pigs and the pigs designed by the genetic engineer with a Lego-like construction system. The British author Brian Stapleford[2] dreams of design chickens without the unnnecessary heads, tails and wings. The food supply and excretions could pass through tubes. What would be the needs of such a chicken torso? What potentialities could be suppressed? It could be considered an ethical progress if bio-engineering succeeded in producing only the economically useful parts of animals. At present the meat we eat reminds us of an absent referent: the living animal and its suffering in the meat factory, its torments during transport to the slaughterhouse, its agony as it is driven inside and dies a violent death that is not always painless. If we were able to produce meat without this absent referent, we could enjoy it without suffering from a bad conscience. The steak at the butcher's would then refer to nothing but itself. It would be an honest part of the world of consumer goods.

Ever since the neolithic revolution we have failed to come to terms with our older brothers, as Johann G. Herder termed the animals. Man was man long before, and as such has consciously distinguished himself from the animals. But man, animal, plant, river, mountain, sun, moon and the stars all 'lived' together in the same democratic household, the household of the Great Tribe. The transition to agriculture and cattle-breeding, the construction of cities and temples, the birth of the patriarchal central state all mark the beginning not only of the opposition between human culture and non-human nature but also of the perilous and potentially deadly adventure of mankind on its own. This adventure seemingly ends with the complete assimilation of non-human nature by human culture. The moment of utmost alienation of man from nature is the very moment where the difference is effaced. The human household has devoured everything outside itself, has included everything in itself. Man's urge for expansion and lust for land and resources seem as insatiable as his unlimited hunger for meat. Being part of the 'first world', being rich and prosperous means: cars, TVs, VCRs, holiday travels and meat. 'Meat is life', the meat industry ads claim. Or: 'Meat is a true force of nature.' But not: 'Meat is a rotting piece of a corpse.'

'There are currently 1.28 billion cattle populating the earth. They take up nearly 24 per cent of the land mass of the planet and consume enough grain to feed hundreds of millions of people. Their combined weight exceeds that of the human population on earth.'[4] More than 70 per cent of the American grain crop and about one third of the global grain crop is used as cattle feed. The worldwide expansion of cattle-breeding is one of the main causes of environmental pollution. It is always the same in every field of modern industrial life: metastatic growth and an unlimited, unbridled and addictive desire. And why not? Why not eat as much meat as our body, aided by modern medicine, will take? Why not have two or three billion cattle on earth? What else do we need modern sciences and technology for? Let's build even bigger meat factories, ten floors high, and pile up even more animals. Let's develop power stations that can convert manure. Or cars that use manure for combustion. Biotechnology for its part should be able to produce all the necessary ingredients for feed. Even our own vegetable food will be affected by increasing biotechnological manipulation. We are in fact on the verge of the complete manipulation and industrialization of life. The consequences are incalculable. The most fundamental categories of being will be affected and uprooted; the distinction between a living organism and a machine will be blurred, we will have to deal with artificial plants, vegetable animals and all sorts of fantastic creatures. Even the distinction between man and animal may fade away: already human genes have been transplanted into the genome of tobacco plants, mice and pigs.

A huge billboard in the city says 'Insect Killer' and shows a spray can containing a deadly liquid capable of exterminating any kind of insect. Albert Schweitzer holds that all life is sacred, including that of insects. 'Numbness is the biggest enemy of ethics',[5] the indifference that allows us to open a window on a summer night without considering how many insects our lamps will burn. This ethical position seems to be unworldly and unrealistic. We bear no grudges against insects, as long as they stay outside our homes and make themselves useful in our gardens. They should not appear in large numbers. And as many of them will not abide by these rules, we have to exterminate them in self-defence, using the weapons provided by the chemical industry. Before these chemical weapons are marketed, they are scrupulously tested to eliminate any potential dangers to man. To this aim the industry needs laboratory animals that are bred in industrial firms. Laboratory animals are also needed for the gentechnological modification experiments; these modified animals are then used for medical and other research. Like the onco-mouse, the mouse that was doomed by genetical manipulation to develop cancer cells.

The ethical implications of animal experiments have since long been the object of fierce and passionate debate, more than any other form of animal exploitation or destruction. The stable, even in the form of the animal factory, is still generally considered to be a fairly adequate living-place for domestic animals; not so the laboratory cage. The breeders of laboratory animals and the scientists using animal experiments are on the defensive, but still not willing to surrender. Their defence policy is to claim that medical progress will halt without animal experiments and that we will be prey to welfare diseases, to cancer, Aids and horrible hereditary diseases. Without animal experiments we will be forced to test product safety on humans. Without animal experiments there will be no more progress in bio-sciences and in psychology. But, they claim, we are willing to reduce the number of animal experiments, to use (or to abuse) more rats and mice instead of runaway or stolen cats and dogs. The priests of the animal experiments like to refer to 'sacrificing' animals.

Stock-breeding animals, laboratory animals and vermin: those are the three species of a zoology determined by human self-interest. These species require an objective, cool and rational approach; no sentimental loose talk. In the case of laboratory animals, this leads to a tell-tale paradox: the projection of laboratory animals to man is defended on the basis of the similarities between man and animal, but the ethical justification of animal experiments invokes an unbridgeable gap between man and animal.

Vermin are a crawling mass that has to be checked with fire and poison — but can never be got under control completely. The rat became the totem of the nihilistic punk culture: the rat as parasitic negation of our well-mannered civilization, gnawing its way into aseptic computer rooms, surviving on islands wiped clear by radiation from nuclear testing. After the last piece of non-human nature has been conquered, after the last spot on earth has been put to a human use, be it as residential area, industrial area, road, recreation area or natural reserve, when the earth's biosphere is managed on a global scale using satellites, countless measurement and control systems and worldwide information and computer networks, 'vermin' will still subversively resist human authority, and may live to be the last testimony of the creativity and fertility of the old, i.e. pre- and non-human, nature.

The animal as slave, the animal as machine, the animal as laboratory animal, the animal as enemy: an unimaginable number of animals are subject to merciless and seemingly casual atrocities, to exploitation, mutilation, destruction and extermination on an industrial scale, in the name of man and his desires. A large part of the material basis of our industrial society rests on an immense system of production and destruction of animal life. But, paradoxically, love of animals and loving interest in animals are growing while the industrial system of animal production and destruction is expanding in the basements of the human household, while the war on our animal enemies is being con-

*Back street with cuddle bear (Teun Voeten, Haarlem, April 1988)*

ducted more and more aggressively. Animal books, animal films, zoos and safari parks: there is a wish to see natural animals, to be fascinated by their similarity to us and by their radical otherness. An intense desire exists to live in a non-instrumental relation to animals. The feeling of being related to and connected with animals, the astonishment at the mystery of animal existence are firmly rooted in our human nature, as can clearly be observed with children. These feelings lie at the basis of the process of human self-definition and development.

But industrial man's love of animals is a narcissistic, possessive, unreliable and often painful or deadly love. There are some 170,000 amateur anglers in Flanders, all of whom probably consider themselves to be lovers of nature and animals. Angling is like the poor man's hunting. Fish have to be planted in large numbers to meet the demand. A special industry has developed to provide the necessary angling equipment, which of course keeps up with technological progress.

The truly masculine love of animals, devoid of sentiment, manifests itself in hunting. The editor of *De Vlaamse Jager* (The Flemish Hunter) writes:

Hunting is not just shooting but a lot more. (...) Hunting is always being in the open country, working the fields, securing nests, keeping track of game stocks, training a dog, paying the land owners, repairing ferret cages, controlling other game, helping the farmers, combatting poaching, watching and listening and being amazed. Hunting is also enjoying a strong dove soup, rabbit's hindquarters, pheasant with a glass of beer, saddle of hare and good wine in crystal glasses, gnawing a young duck, together with friends. Hunting is also the feeling of the hunter that lies beyond words, the feeling right before and after the gunshot. The shot creates the hunter and the hunted, brings them together, joins them, and makes life and death a vital part of the hunt.[6]

The same applies to the fields and meadows near the village where I live, as to the surroundings of Tarascon: 'As for the game, as for the small animals with fur or wings and feathers, they are in a bad way.'[7] Thus the masculine hunters are forced to release young pheasant and partridges, which are supposed to develop into 'game' within two months. The rest of the game consists of an occasional magpie but mainly of cats — which are held under control by poisoned bait. But a hunter's destiny is to hunt; just as the destiny of game (according to *De Vlaamse Jager*) is to lose its life in the mystic union of the hunt. But the only blood shed comes from animals riddled with lead shot.

A form of animal love totally different from the one displayed by the hunter is the lasting personal bond, an I-thou relation, with an individual animal. From the anonymous mass of animals man singles out individual animals, calls them by a personal name and accepts them in his home as family members. The animal-protection law terms these chosen animals 'animal companions'. They seem to lead privileged lives compared to their relatives in stables and laboratories. They enjoy personal attention and care from almighty man. The enormous number of animal companions illustrates the great need felt by man to live in the company of animals. Belgium alone is home to three million cats and dogs, plus all budgies, rabbits, hamsters, guinea-pigs and other animals treated as family members. But the fate of many of the millions of animal companions can hardly be envied. Each year in Belgium about 250,000 dogs and cats are dumped into animal shelters. And precisely those who fight with heart and soul for animal rights see themselves forced to handle this endless row of abandoned animals cast from human company by wielding a deadly needle.

The wants and whims of the master or mistress determine the terms of the contract with animal companions. They have to adapt themselves, submit, obey, be docile, live up to expectations — no matter whether these are compatible with their disposition or individual character. The animal as pet, as toy, as servant, as object of prestige, as child, as clown, as life companion, but only seldom the animal as itself, with respect for its individuality. Once the genetic manipulation of animals has become a respectable and accurate technology, it will certainly be possible to buy animal companions made to order, who can easily live up to our wishes and expectations.

Few animals larger than insects still manage to live in our industrialized concrete jungle without attracting man's attention, i.e. without being hunted, caught, controlled or exterminated and being threatened as a species. They include mainly birds and some smaller mammals like the squirrel and the hedgehog — who seem destined to be crushed under the wheels of a car. The numbers of animal road victims are countless and uncounted.

The world is becoming too small for free animals. The world population is steadily growing, more and more people live with one desire: to enjoy the same material prosperity as the average Westerner. Material prosperity implies an elaborate system of animal production, so the animal population that is included in and forms part of the industrial production process also grows. Wolfram Ziegler has calculated the weight per hectare of the humans, the animals kept and exploited by us, and all other animals including birds, in the old Federal Republic of Germany. According to his calculations, we weigh 150 kg/ha, the anthropogenous animals 300 kg/ha and the other animals 8 to 8.5 kg/ha.[8] Is there still a future on this planet for the wild, undomesticated animals, the animals that are not yet part of human culture, or will only a shadow of them live on as a tourist attraction in wildlife reserves and zoos?

What are things coming to — for us and for the animals? The majority of people do not seem to have any scruples regarding the enormous animal suffering for which they are directly or indirectly responsible. Is this just indifference or lack of knowledge? Do most people ignore what is done in their name? Is this moral blindness or moral arrogance? — Animals are not worthy of our moral attention or effort. Even if we are biologically related, we have by far surpassed them in our cultural evolution. Are these deeply rooted religious convictions? — Only man has a soul; only man was given the freedom to choose between good and evil, so that only man is responsible to God; only man has to rely on and can hope for the mercy of God; Christ has died for mankind only, etcetera. Or is it just moral common sense? — We have enough trouble stopping the cruelties of man to man.

Our behaviour, our attitudes towards animals in all their complexity and contradictions above all are culturally and socially self-evident. Starting with the first chunks of meat fed to ten-month old babies, we assimilate the naturalness of cattle and of the animal production and destruction system. But some signs indicate that this naturalness is questioned. A worldwide, decisive and determined animal liberation and animal rights movement is breaking silence on the horrors of animal factories and laboratories. It demands that the moral and legal rights of the individual animal be recognized: to live according to its nature, free from exploitation and human arbitrariness. Under the influence of increasing public susceptibility to animal suffering, the laws on animal treatment and animal protection have been changed in favour of the animal in most Western countries. The sensitivity to the unseen source of our food — the animals, the Third-World farmers, the agricultural industry — is also increasing. The growing concern for wholesome, non-violent and fair food stimulates vegetarianism. The enormous animal production system is exposed by the ecological movement as one of the most pollutive and destructive factors in our environment. The stress here lies on preserving the natural ecosystem in which free, natural animals can live. More and more we grow aware that the comfort of the coming post-natural world with its post-natural plants and animals does not outweigh the loss of old nature.

As yet one can not speak of a revolutionary change of pattern in our relation to animals, only of a heightened susceptibility to the fact that what is done to animals by us and in our name is not inessential but deserves the attention of the modern, pragmatic and rational man. The abuse and humiliation of animals, the denial of their right to exist according to their own natural norm and 'telos', the lack of respect for their otherness, for their own beauty and perfection, the denial of their inner self, of their pains and suffering, of their desire for happiness, the arrogant pretence that they exist for us and not for themselves, and finally the extreme logic of domestication, the arrogance of replacing the old animals by high-tech animals: is it not possible that all this is unfair to ourselves, to our humanity and our evolution as humans? Could it be that violence, like justice, is indivisible? Could it be that we will only find a way out from the spiral of destruction in which the industrial system of modern times has brought mankind if we begin to look at animals in a different way? Could it be that their and our fate are linked in other, more fundamental ways than in simple terms of human domination and power?

What are things coming to — for us and for the animals? Do animals need to be 'liberated', to be recognized as legal subjects? Do we have to become vegetarians? Do all household animals, all domesticated animals have to disappear in the long term? Do we have to fence off and guard our roads to prevent animals from falling victim to traffic? Do we have to provide hospitals and rest homes for animals? The precise character of our mutual relations in future still is unclear. The concrete utopia of a culture which, in Albert Schweitzer's words, maintains 'a good and fair relation to animals' is as yet unformed.[9] But the guiding principle for our future behaviour to our fellow creatures has been admirably formulated by the great English pacifist and humanist Henry S. Salt: 'No human being is justified in regarding any animal whatsoever as a meaningless automaton, to be worked, or tortured, or eaten, as the case may be, for the mere object of satisfying the wants and whims of mankind.'[10]

*Translated from the Dutch by Eric de Rijcke*

## Notes

[1] H.-D. Dannenberg, *Schwein haben. Historisches und Histörchen vom Schwein.* (G. Fischer Verlag: Jena 1990), pp. 162 ff.

[2] B. McKibben, *The End of Nature* (Penguin Books: London 1990), p. 151.

[3] On the absent referent, see the impressive study by the feminist author Carol J. Adams, *The Sexual Politics of Meat* (Polity Press: Cambridge 1990), which traces back the link between patriarchism and meat on the one hand and feminism and vegetarianism on the other.

[4] J. Rifkin, *Beyond Beef, the Rise and Fall of Cattle Culture* ( Penguin: New York/London 1992), p. 1.

[5] A. Schweitzer, *Eerbied voor het leven. Hoofdstukken uit zijn ethiek, vertaald en samengesteld door Hans Bouma* (J.M. Voorhoeve: Den Haag s.d.), p. 50.

[6] *De Vlaamse Jager*, June 1988.

[7] A. Daudet, *Die wunderbare Abenteuer des Herrn Tartarin aus Tarascon* ( Philipp Reclam: Stuttgart: June 1968), p. 7.

[8] Cf. R. Bahro, *Die Logik der Rettung. Wer kann die Apokalypse aufhalten? Ein Versuch über die Grundlagen ökologischer Politik* (Weitbrecht: Stuttgart/Wien 1987), p. 30.

[9] A. Schweitzer, *op.cit.*, p. 22.

[10] H.S. Salt, *Animal Rights, Considered in Relation to Social Progress* (Centaur Press: London 1980), pp. 16 ff. [1st edn 1892].

# When the King learns to do it his own way: animals in the age of Enlightenment

## Paul Pelckmans

Nowadays historians are happy to investigate how even the most trivial and apparently unvarying commonplaces throughout the centuries have not been at all so unchanging. The history of attitudes explores unsuspected developments in banal universals. To my surprise, however, I must say that the way in which people have dealt with animals seldom comes under discussion here. The few historians who write about animals talk about the changing composition of the livestock and such down-to-earth topics. But an anthropology, a historical psychology of the relationship between man and animal, has to the best of my knowledge not yet been written. If the theme is put into a cultural index, it is usually in a summary semiotic, which spreads the animals out over a field of exclusively human significance. Anyone who ignores the traditional, mystical or moral allegories opts for a psychoanalytical interpretation: that seems more modern, but ultimately it makes no difference, since mental images are still human-subjective.

I would like to suggest here that the relationship between man and animal has a history all of its own. I shall restrict myself to a couple of random indications.

The time between the *Ancien Régime* and the Enlightenment is for historians of attitude a decisive turning-point: a series of now classical studies on very different topics has uncovered a fundamental loss of certainty, a complete disruption of all the obvious relationships and involvements up to that time. My question is whether that disorder points to a new attitude by the philosophers of the Enlightenment and their readers towards animals. Do people nowadays feel themselves less directly involved with animals than their ancestors? There is not a single historian who will contend that a sharply defined division arose between the pastoral-idyllic *good old days* and modern indifference. But perhaps we could also look at the change of attitude another way.

The *Ancien Régime* seemed to involve all sorts of popular entertainments which to our taste were brutal; the crude enjoyment of the goose riders bears witness not so much to familiarity with matters spiritual, but more to solid indifference. But it suited a time when even the farmers used to leave their animals out for months on end[1] and the abattoir was to be found in the middle of town: the neighbours complained about the smell, but saw no reason to worry that the bloody work was carried on largely in the street. That selfsame indifference explains, in a completely different context, the obvious blunders in the *Fables* of Jean de La Fontaine. Crickets and ants do not eat grain, foxes and ravens are not fond of cheese; the fable-teller did not often have to concern himself in such matters: the townsfolk and others who read his *Fables* encountered so many animals in their daily lives that they no longer bothered about it. Emotions first come to the surface when relationships become rarer and less obvious. Then they receive — sometimes — more emphasis. In 1907 Daniel Halévy closed the first of his famous reports on *Les Paysans du Centre* with a conversation in a well-run stable. The host's younger brother, who had accompanied the visit without saying a word, unexpectedly came out of the corner with a sort of oracle:

'Ah,' he said to me in a subdued tone which has remained in my memory to this day 'living with animals really isn't bad...' 'Of course,' called his elder brother, 'no farmer would ever leave for the city if life here was not so difficult!' Guillaumin looked at me in silence, as if he wanted to say: 'Now just you listen to those two guys!' I didn't need him to tell me to take extra care.[2]

The young man's statement did not come over very well: without the 'accent' and the italics, people would just read through it without getting the point. The older brother does step in hurriedly, but probably has more tangible things to say about the discomforts of living in the country. The peculiar response of the only positive note in the conversation principally shows in this respect that contentment-with-animals in 1907 was markedly less obvious.

## A natural history?

To learn how things taken for granted began to be questioned, I shall start with the most important animal book of the Enlightenment — Comte de Buffon's *Histoire Naturelle* (1749-67). Buffon positively loved animals and, to demonstrate his enthusiasm, found more detailed and convincing words than the non-committal negation which Halévy had to make do with, a little less than a century and a half later. This time the response was European.[3] For the history of attitudes a success of that magnitude signifies a sort of plebiscite: Buffon apparently orchestrated a vision of nature in which his enlightened readers recognized themselves.

The seventeen weighty tomes offer primarily a treasury of practical information. This was not entirely new. Modern anthologies usually give a distorted view of the *Histoire Naturelle*: they mainly quote from the visionary pages, where Buffon anticipates later developments in natural science. But his text contains at least the same amount of traditional material. He talks at great length about breeding horses, fattening up pigs or raising hunting dogs; the little chapters on wild animals always contain some advice for hunters. Many critics of the Enlightenment assert — in fact since the French Revolution — that the *Philosophes* concerned themselves with abstractions, with general and thus useless discussions of principles, which in practice were even dangerous. That criticism will sound rather insubstantial when you see how comprehensive that bible of the Enlightenment stands with the know-how which has grown up through the centuries in kennels and stables. In the middle of the eighteenth century that was nothing special: The *Encyclopédie* won over its readers with a comparable mixture of daring 'philosophical' polemical articles and more down-to-earth contributions on the old handicrafts and the first machines. What was new, was that the facts were collected in inclusive anthologies. The falconers in Versailles most probably did not know which breed of pig provided the best bacon; their dog-breeding col-

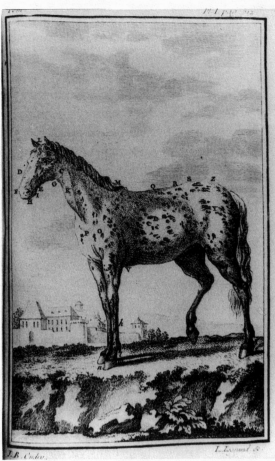

*The young wild boar and suckling-pig, and the horse, from Buffon,* Histoire Naturelle *(1749-67)*

leagues had nothing to do with the world of the cheese-maker. The *Histoire Naturelle* describes traditions which previously had been handed down separately and in closed circles; collecting them together makes them unusually impressive. On the surface of it, Buffon does not offer any new advice, but because his counsels are set out in an endless series one after the other, they suggest unlimited power over nature. People in the eighteenth century were not 'capable' of being any more objective than their forefathers, but the *Histoire Naturelle* proposes a new subjectivity, a claim to omnipotence which only generations later was to come true in genetic and other laboratories. For that typically modern presumption, Buffon was not able to call upon any real achievements of his own yet; he substantiates them with a parade of old skills dispersed hither and thither.

That such limitless ambition was to unsettle relations with animals is obvious. The *Histoire Naturelle* claims the diversity of domestic animals as the greatest triumph of human ingenuity. Breeders succeeded again and again in creating new breeds through patient crossing of exceptional or differing species — which were then more efficient than their predecessors! First, in the foreword to the volume on domestic animals, it is still reported that people knew how to vary the individuals but not the species, but such limits are soon forgotten, the impact of Man on animal species then actually becomes 'the most beautiful right over nature that he has'. At the same time there are a number of passages which seem dismissive of that interference with the animals:

All animals in slavery bear the scars of their imprisonment, the marks of their chains. Just compare our little sheep with the moufflon, from which they were originally bred. As large and agile as a deer, equipped with defensive horns and heavy hooves, wrapped up in a thick fleece, the moufflon has nothing to fear from the ravenous wolf and can withstand inclement weather; not only can he escape from his enemies because of his speed on the hoof, but through his power and the heavy weapons with which his head and feet are equipped, he can also fight them back. And just look at our little sheep — their skins are no match for our Winters: in short, they would all die if Man did not look after and protect them continuously.[4]

Buffon worked for twenty years on his *Histoire Naturelle,* but that does not mean that he changed his viewpoint along the way. Throughout his entire *oeuvre*, he talks with understandable pride of the innumerable, ingenious ways in which Man put the animal kingdom to his own use. Even in the remarkable epilogue entitled 'De la dégénération des animaux', in which there is a predominantly pessimistic note, he enthusiastically suggests new hybrids, to improve the sheep and the goat. If he once regarded such interference more negatively, that only appears as an occasionally occurring scruple, a feeling of guilt which never gets any further than moralizing against unnecessary — never against obviously painful[5] — interference. In addition, he regularly involves himself in some kind of theatrical striving for effect in his discussions on the original glories of the animal kingdom before Man cut off the pass to all other species. Since the seventeenth century travellers in Canada have been writing enthusiastic reports of the ingenious dams and dwellings on stilts built by the beavers; Buffon suggests that they, in their impenetrable corner of the world, had perhaps managed to safeguard something of an animal ingenuity elsewhere suppressed ...

The *Histoire Naturelle* appears to be permeated with a certain hesitancy, a reserve which never really gets the upper hand but which at the same time never really disappears.[6] This ambivalence can be explained, I think, by the fact that the human omnipotence claimed by Buffon presupposes an absolute freedom of movement. It presupposes an *instrumental* or *artificialized* approach to nature. Two centuries later we are no longer surprised by this, but anthropologically speaking this is absolutely not commonplace. Ultimately — the analyses are sufficiently well-known — the unlimited right to try and rearrange everything is a corollary of modern individualism: in traditional societies it was unthinkable that everyone could go his own way unhindered. The Enlightenment fought for that unheard-of freedom, which put an end to all restricting relationships and horizons. For the animals it was Descartes who brought an end to them, he who coldly spoke of *'animaux-machines'*. Buffon was regularly at loggerheads with Descartes, but spontaneously took up the view held by him that (prehistoric) man was felicitously inspired when he got the brilliant idea of breeding dogs:

The machine, these instruments we have invented to perfect our senses, to give them greater range, cannot — as far as usefulness goes — hold a candle to the machines that nature has provided us with: they compensate for what is lacking in our sense of smell and thus provide us with enduring, powerful means with which to conquer and control.[7]

It remains in the occasional reminiscence. The image of the animal-machine fits in better with a philosophical deduction than in the detailed evocations of the *Histoire Naturelle*. Anyone who looks at the animals closely will inevitably see subjectivity and directed movement. The animal is in its own way capable of individual freewill which becomes, with the Enlightenment, a universal right. Buffon clearly does not conclude that they enjoy the same unconditional right; it does cause him some remorse for the reason that the unfettered

LE LEVRIER.

*The greyhound, from Buffon,* Histoire Naturelle *(1749-67)*

life of the 'wild' animals fits in too perfectly with his individualistic desires: 'Love and freedom, what blessings! The animals we call wild because they are not in our power, what more do they need to be happy? Among themselves they are equal; they are neither the slaves nor the tyrants of their species ...'[8]

The most famous description of the *Histoire Naturelle* presents the tamed horse as a triumphant success, 'the most noble conquest ever achieved by Man'. But a little later the wild horses of the Pampas seem to be more impressive:

Nature is more beautiful than Art, in a living being it is the freedom of its movements which Nature makes so beautiful. Just look at the horses which have multiplied in the Spanish Americas and live there as free horses. Their walking, their galloping, their jumping are never hindered or controlled. Proud of their independence, they avoid the company of Man, they spurn his cares. They seek out and find the food they need themselves. In freedom do they roam and run through the interminable Pampas ...[9]

Where human omnipotence reached sublime despotism, it had sometimes to feel guilty.


*When the King learns to do things his own way*
Modern individualism,[10] as opposed to the freedom of the animals, falls into a kind of impasse. Lifeless nature is something that men can deal with without any moral difficulties: but for rare exceptions, the Enlightenment still had no ecological cares.[11] However, in accordance with the famous ethical constitution of the *Kritik der praktischen Vernunft* (always an aspiration), fellow-man could never be a means to an end. The semi-subjectivity of animals fell between these two stools: freedom, which ceases where that of another begins however, clashes here with an irritatingly changeable boundary. Buffon was therefore not entirely successful in his bid to identify fundamental activism. Where that activism is absent, men — even in the eighteenth century — listen to a more traditional voice: anthropological movements are rarely radical mutations. The fashionable Abbé Bougeant, for example, published in 1739 — when Buffon was making the first notes for his *Histoire Naturelle* — an *Amusement philosophique sur le langage des bêtes*. The little treatise latches on to a distant sequel to Locke's critique of Descartes. Bougeant popularizes the debate and sets out a series of

traditional and less usual examples for and against. His final argument winds up with an appeal to common sense:

I shall restrict myself to an important argument, which in my opinion is as good as proof. Every day we speak to animals and they understand us very well. The shepherd makes himself understood to his sheep, the cows understand everything the milkmaids say to them; we speak with our horses, our dogs, our birds and they understand us. For their part, the animals speak to us and we understand them.[12]

I suppose that Abbé Bougeant was entertaining his lady readers with popular, straightforward language. It shows at least the view shared at the time which was apparently widely held.

I will not dwell here on the attitude of taking things for granted which made that easy intercourse possible. We shall, as so often in the history of attitudes, touch on a psychological *Ancien Régime* which we can still barely imagine. Traditional societies, in which all interested parties took part in their general evolution, used their animals without asking difficult questions: there was no need for scruples, though everyone was obliged to make his own particular contribution. Nor were there any really disturbing interventions: the initiatives remained contained by all manner of established practices and regulations.

Plans and projects seemed less important than the stability of the whole, a whole in which animals had their rightful place next to all the rest.

The art historian Emile Male used to remark that the carved fauna in gothic cathedrals concealed fewer allegorical themes than might be imagined.[13] In guided tours the public are overwhelmed with allegorical explanations: the modern tourist, however, finds it obvious that their creators had to have something to work on. In reality, they chose and grouped most of their sandstone animals without any particular reason. The underlying theology limited itself to a general conviction that all God's creatures belonged in the Church — *il faut de tout pour faire un monde*. For the rest, it is not unknown for some animals indeed to represent the four evangelists or the cardinal virtues: in a fully coherent world artists invented a few extra relationships without any problem.

*Pets*
The old conviviality we see in the cathedrals jars against modern initiatives. In

his reference work on the English evolution, Keith Thomas presents us with another scenario. He sets the eighteenth-century scruples off against an instrumental attitude which would be as old as humanity itself. However, civilization always resided in the control or exploitation of nature. With all respect for a study which of course is much more fully documented than my brief sketch, I find that Thomas' vision disguises an essential discontinuity: the traditional use of nature remained much more discreet than the unlimited interventions of our modern age. To be sure, our Western tradition appeals to a biblical kingship over creation, but that occurred in a time when even actual kings felt themselves bound by centuries-old customs and unwritten laws. Modern free-choice only begins when the King — and then everyone — learns to do things completely on his own.

Keith Thomas quotes quite a lot of material which in my view points in a more convivial direction. He gives us a long introduction to the eighteenth century from the middle ages or even from a number of animal-friendly texts in the Bible. To the question of why the counter-current took so long to get off the ground we get no definitive answer. The lines are actually rather sketchily drawn: in one breath, in the same chapter, he discusses the traditional familiarity of farmers and herdsmen with their animals and the very differently inclined modern success with pets ...[14]

At first sight, that success does not accord with the triumph of individualism either. Pets are by definition not treated instrumentally. If one looks up the literary allusions there, one must ask the question whether the contrast really is so radical. The landlady who tells the well-known story about the Marquise de la Pommeraye in *Jacques le Fataliste* is probably the oldest French character in a novel who is expressly typified by her passion for her lapdog. In a later conversation Jacques makes a comment that sounds unexpectedly individualistic:

Jacques asked his master if it had not struck him that all paupers, however poor they were, though they might not have enough to eat themselves, almost all still kept a dog. Had he not noticed that those dogs were all trained to go round in circles, to walk on their hind legs, to dance, to bring back a thrown stick, to make little jumps for the King or Queen, to lie down and pretend to be dead? Did he not realize that those training sessions had made those animals the most unhappy animals in the whole world? Jacques deduced from this that all men wanted to rule other men, and given that animals were to be found on the social ladder immediately below the class of the lowest placed people ruled over by all other classes, paupers kept an animal so that they too would have someone to rule over.[15]

The comment is all the more revealing because, within the plot, it literally makes no sense at all. The landlady's little bitch doesn't have to perform any tricks, she is not ordered about at all, but apparently only cuddled and stroked; in affection Diderot discovers some free choice for poor people.

Sébastien Mercier traces the line in the 670th chapter of his enormous *Tableau de Paris*. The poorer the Parisians are, the more animals they keep; the attics and tenements are home to a surprising collection of fauna. Mercier begins his story with the rabbits, who — thanks to their proverbial fertility — provide their owners with cheap meat on holidays. The following paragraphs deal with less pragmatic motivations:

Tailors, shoemakers, precious metal workers, embroiderers, seamstresses — all the sedentary professions always keep some kind of animal in a cage, as if to share their own slavery. A magpie for example, who does nothing throughout her entire life in the cage except fly back and forth in the hope of finally finding a way out. The tailor looks at her and decrees that she must keep him company for ever.[16]

The housebound all have their pets. Mercier might have commented that sociability requires companionship; he prefers to write about a grudging need to impose imprisonment on a helpless victim. Tame birds which would feel at home in their cages might just as well be locked out: they just fly around in the vain hope of ever escaping. The magpie becomes such a prisoner, even doomed: 'for ever' in the *Ancien Régime* refers to a well-defined maximum sentence, to which the tailor in Mercier's story by royal command — 'and decrees' — condemns his magpie. People who keep larger animals enjoy greater freedom of choice because they can make life unpleasant for their neighbours: a barking dog or a flapping magpie can at the same time disturb the peace of the entire neighbourhood.[17]

The sentimental novel of course sounds more warm-hearted. Only with re-reading would you suspect that the expressly displayed emotion competes with an aloofness which is never quite overcome. Sterne's *Sentimental Journey* is one of the most enjoyable specimens of the genre. Its charm rests principally in a series of casual comments, where you never know for sure whether they are naïve mistakes or discreet bits of deconstruction. In a godforsaken inn between Montreuil and Amiens the traveller Yorick meets an old German farmer, who is weeping bitterly for his dead donkey. This time the grief is not demonstrative; the donkey is for the farmer a plebeian companion, in that sense more believable than a pet which is always somewhat affected. The emotion is thus purer, it comes close to the old conviviality,[18] the difference is only that in the earlier holistic perspective there was barely room for the pathetic-exclusive relationship which the donkey-companion here makes irreplaceable for this farmer.[19] The euphoria would be complete if Sterne did not end by lamenting that such a deep bond is still difficult to find these days: 'Let the world be ashamed of itself, I said to myself; if we could but love each other as that poor soul loved his donkey, what a thing that would be!'[20]

The tears of the farmer for his donkey are like the widow's mite of the Gospels: they prove above all how very inadequate emotion is elsewhere. The more conventional sentimental novel cloaks that shame — and others still — with the mantle of tenderness.

### 'L'homme, mes frères, est un ridicule animal'
The promotion of the pet orchestrates freedom of choice and averts aloofness. We remain within the grip of modern individualism. The headstrong exploitation of which Buffon was the great, though sometimes tormented, advocate is a direct exponent of this; the success of pet animals is also evidence of a sort of compensation. To end my brief *Rundschau*, a curiosity: the philosophers of the Enlightenment spouted forth their social critiques through the mouths of external characters; visitors and strangers note more quickly than 'natives' that what is accepted as normal can be quite insane. Montesquieu wrote his *Lettres persanes* about the visit of a group of prominent Orientals to Paris. The scene was repeated with Incas, Iroquois, Negroes and even, in Voltaire's *Micromegas,* with an astronaut from Sirius. The prolific writer Restif de la Bretonne allows those critical themes to be expressed by an ape, César van Malacca, who tries to convince his fellow species that they should not regard humans with such admiration: 'L'homme, mes frères, l'homme qui nous domine tous, est un ridicule animal.'[21]

*The monkey, from Buffon,* Histoire Naturelle *(1749-67)*

The *Histoire Naturelle* has not a good word to say for apes. The species is said to be incomparably more stupid than for example the dog or the elephant, they only attract attention because a considerable anatomical similarity makes it possible for them literally to ape humans. Buffon was quite annoyed by this: docile imitation is about the lowest a true individualist could imagine himself doing. César van Malacca is at that point which is beyond all suspicion: he calls on his fellow animals never to abandon their instinctive happiness for the dubious blessings of human civilization. Restif's considerations speak about an animal happiness that would otherwise seem to have the individualist's signature.

If you leave free love, an eternal dream, out of the argument, that happiness is limited to two advantages. Animals are primordially happier because they do not envisage unhappiness and in particular have no knowledge of death:

Whoever knows not death, as I once did and you still do, whoever knows and feels only the present moment — he is actually immortal. (...) Humans know that they must die; everything reminds them of that knowledge, which spoils all their pleasures. They sometimes try to forget it, but in reality they are doing nothing other than trying their best to return to the state of innocence which on one wretched day they forfeited![22]

César walks headfirst into a death taboo that for most psychoanalysts draws attention to a particular impotence for the modern individual: the person who sets himself apart, experiences his own end, which then becomes an absolute catastrophe, as a personal humiliation. Because there is nothing one can do about it,[23] elimination is the only, fragile solution. On the other hand, with animals the disaster lurks below the horizon. 'The humiliation, the irritation, the shame, the hate'[24] which belong to the prospect of death do not occur in their case; these terms prove that it was less a question of the fear of death than of the much more specific inconveniences of humiliation and resentment. The animals escape to yet another resentment. Humans are used to eying each other up and depend constantly on the appreciation and goodwill of their fellows:

What they feel, they feel both outside and within themselves. They leave nothing untouched, but their worst torment lies in their envy of another's happiness, which they want to rule over, to raise themselves above the others; it lies in the fact that they suffer because of it, if they are dominated by others, oppressed or trapped. That sometimes brings them literally to a frenzy ... What a difference between our natural state and theirs![25]

All things considered, it comes down to the fact that Restif dismisses any form of inter-human involvement. The animals need no admiration and are thus exclusively happy. The idea that the wise man derives his happiness from himself is as old as the Stoics; traditionally, it was used as a counter to the whims of Fortune. César invokes it with continually disagreeable fellows. The *Lettre* has the necessary criticism of human tyranny over nature; it is supported by the individualism that gave those interventions a new zest. Not surprisingly, convivial familiarity is not reinstated. Restif was the son of a gentleman farmer in Burgundy and was happy to put his national roots in the picture; he might also have put words into the mouth of one of the sheep or guard dogs he used to accompany daily when he was a child — but an ape like César van Malacca with his quasi-human appearance is a more likely letter-writer. César is after all an unusual specimen, the result of cross-breeding between a baboon and a negress; moreover, he was brought up as the favourite pet of an enlightened countess. Of the directness with which La Fontaine put words into the mouths of his animals there is no question. The *Lettre* is not even a real communication. César writes the letter for himself as an aid to memory for the message he wants to pass on to his fellow apes, none of whom could read. In the 'Avis de l'éditeur' we learn how even that was an illusion: 'We owe this remarkable piece to a mistake: César thought that he would be able to make himself understood to other apes.'[26]

César himself tells how he first wanted to give some advice to his mistress's domestic cat. Here too he was unsuccessful: he was not yet intelligent enough to be able to descend to her level!

Restif's bizarre complications are not of course aberrant. César van Malacca is not a victim of the abuse of power: neither does the *Lettre* deal with his disguise role as pet animal. The ape principally discusses the foolish obstacles humans tend to put in the way of their own freedom. When the argument becomes just a story, he too must experience for himself that she inevitably isolates those involved with her.

*Translated from the Dutch by Milt Papatheophanes*

## Notes and references

[1] See Ferdinand Braudel, *L'Identité de la France III* (Arthaud: Paris), pp.75-9 ('L'élevage ancient. Première règle: aux animaux de se débrouiller pour vivre').

[2] Daniel Halévy, *Visites aux paysans du Centre* (Grasset: Paris 1978), pp.96-7.

[3] 'Le succès était 'prodigieux'. Il devait se maintenir tout au long de la publication de l'oeuvre, et l'on sait que *l'Histoire naturelle* fut l'ouvrage le plus répandu du XVIIIème siècle, battant le *Spectacle de la Nature* de l'Abbé Pluch, *l'Encyclopédie* de Diderot et d'Alembert, et même les oeuvres les plus connues de Voltaire et de Rousseau. Buffon voulait toucher le grand public; il y avait pleinement réussi.' (Jacques Roger, *Buffon. Un philosophe au Jardin du Roi* [Fayard: Paris 1989,] p.248.)

[4] *Oeuvres complètes de Buffon* (Garnier: Paris n.d., 2nd edn. M. Flourens [ed.]), tome IV, pp.13-14.

[5] Buffon himself liked to experiment with cross-breeding. He then had animals of possible related species raised together to see if they would indeed mate and whether the foetus was viable. The animals, both of which were isolated from their own species, were not always pleased about it: 'Ce n'était plus que des hurlements de douleurs mêlés à des cris de colère; ils maigrirent tous deux en moins de trois semaines, sans jamais s'approcher autrement que pour se déchirer; enfin, ils s'acharnèrent si fort l'un contre l'autre, que le chien tua la louve, qui était devenue la plus maigre et la plus faible, et l'on fut obligé de tuer le chien quelques jours après, parce qu'au moment qu'on voulut le mettre en liberté il fit un grand dégât en se lançant avec fureur sur les volailles, sur les chiens, et mêmes sur les homme.' (*Ib.* t. II, p.487)

[6] On this topic, see Jacques Roger's *Buffon*, *op. cit.*, pp.306-16, 348-51 and 392-8.

[7] *Oeuvres complètes de Buffon*, *op. cit.*, vol. II, p.476.

[8] *Ibid.*, vol.II, pp.505.

[9] *Ibid.*, vol. II, pp.369, 370.

[10] Enthusiastic readers of Buffon will perhaps note that his *Histoire Naturelle* does not have such an individualistic slant. The *Histoire* inevitably devotes itself to the social life of some animals; it was not so obvious that Buffon often ascribes the superiority of Man to the benevolent influence of his sociability. People do not arrive as adults into the world, they must be cared for by adults for years: this lengthy interaction changes his instincts into intellect and his grunts and cries into language. It does not sound immediately individualistic, but serves in fact to finish with another type of genealogy in which Man's special place in nature was expressly designed by his Creator. The creation implies a duty to be grateful, which expires with Buffon, where the new thinking has a splendid but unforeseen consequence, so that his scenario promises a freedom equal to — and even greater than — the contemporary accounts of the 'contrat social'.

[11] Even in the nineteenth century scruples were only expressed over the rough treatment of *domestic* animals. It still seemed incomprehensible that plant species or 'wild' animals could

seriously be under threat. Cf. Maurice Agulhon, 'Le Sang des Bêtes. Le Problème de la Protection des Animaux en France au XIXème siècle', in *Romantisme* 31 (1981), pp.81-109.

[12] Abbé Bougeant, *Amusement philosophique sur le langage des bêtes*, H. Hastings (ed.) (Droz: Geneva 1954), pp.83.

[13] Cf. Emile Male, *L'Art Religieux du XIIIème siècle en France* (1898), (Colin : Paris 1958), pp.73-136 .

[14] Cf. Keith Thomas, *Man and the Natural World: Changing Attitudes in England, 1500-1800* (Penguin: Harmondsworth 1984), p.100-20.

[15] Diderot, *Oeuvres* (Gallimard: Paris 1951), p.618.

[16] Sébastien Mercier, *Tableau de Paris* (1783) (Slatkine: Geneva 1979), vol. 8, p.336.

[17] Chapter CCXLIV of the *Tableau* ('Les petits chiens') in the same way describes 'petites maîtresses' who let their admirers crawl like their lap dogs (cf. vol. 3, pp.133-5).

[18] In this respect it is not without significance that the farmer is returning from a pilgrimage to Compostella: Sterne's illuminating travel story also intersects with a far more archaic relocation.

[19] Cf. Ariès' well-known contrast between *'la mort apprivoisée'* as a sociable and thus acceptable occurrence and *'la mort de toi'* as a permanent separation from an intimate [*L'homme devant la mort* (Seuil: Paris 1977)].

[20] Laurence Sterne, *A Sentimental Journey, the Journal to Eliza* (Dent: London 1962), p.44. Briefly: *A Sentimental Journey* also contains an episode about a robin, who felt as unhappy in his cage as Mercier's magpie. Yorick is deeply moved, but unable to break open the cage. When his squire buys the little creature for him, Yorick takes it to Italy and thence back to England. Now that he has become the owner, he no longer thinks of setting it free.

[21] Restif de la Bretonne, *Lettre d'un singe aux animaux de son espèce* (1781) (Levallois-Perret: Manya 1990), p.35.

[22] *Ibid*, pp. 25-6.

[23] The painful prospect again comes under discussion in the concluding summing-up of the *Lettre*. Here, Restif risks a naughty bit of *'Vereinung'*: 'Une mort tranquille et accidentelle est pressentie par lui dès l'enfance et abreuve de fiel tous ses plaisirs' (*ibid*, p.54). The death of a living being is anything but accidental.

[24] *Ibid.*, p.20.

[25] *Ibid.*, pp.27-8.

[26] *Ibid.*, p.20.

# The borderlines

## *Anne Cauquelin*

I hesitate between text and rhyme. Between the two. Between the extremes. Between one side and the other. Between the ante and the post. Do they have to be opposites? We hunger for borderlines, we search for them, we create them: we need them.

One might wonder where this essential principle of separation originates from. It arose from customs, but how? These classifications, these limits.

To be on the borderline, exactly on the line. 'To stand on the barricades, all right, but on which side?' said someone looking for trouble. An uncomfortable situation, it could almost get you killed and ye ...

So, where does this need come from to define oneself by excluding everything that is not oneself. The classics did not worry about it, or rather resolved the problem of separation in their own way. By staying alert. By making a clear distinction between qualities of the one party (humans) and the other (animals). Not lines which fork and invariably separate, as with Linnaeus, producing universal types, but tangled and always strange shapes, each a little adventure in itself.

*Montages-collages,* just enough to give the dream a cosmic flavour. A comic note. For one must enjoy, laugh and sneer. One too often forgets — because of a sort of profound seriousness which puts one's mind at ease, to the cost of real perception — that antiquity is not the age of the 'great stories', not all the time, not continuously, not in every field. There is a lot of descriptive space round the questions of existence and time: margins, annexes where one can breath. One finds most of these peculiarities in zoology which does not only concern animals but any living being.

Because we are alive. We are transitory, destined to die, to generate and perish. Melancholy? No, enduring.

Large categories? No, sets of enclosed areas in the text of the world.
The separation unites, the cut leads to a sticking back together, this is nothing new. The doors, the pores through which one side and the other, text and rhyme, ante and post combine. Do they have to be either open or closed? To the classics they are swing-doors.

To position ourselves in time, post, ante, trans-post, we relegate the extra-ordinary to fiction. We are lukewarm, we live in the lukewarm. We are satisfied with a poor duck with a cap, a common mouse, crocodiles painted on a pull-over, distorted machine-like frogs. When the imagination is not sustained by the believe in its own existence, it is barren.

By mocking the common belief, we would only be hatching a plot. At best, we can quote it, which is another form of conspiracy. Within fiction, we invent, we imagine. Because we expel the things which bother us to the world of dreams. Separation: we humans here, and there, behind the wall of fiction, the strange and unbearable. We do not believe in it, it just amuses us. The kind of amuse-ment which makes you shiver, on the verge of fear.

The classics do not imagine, do not invent: they describe what they have seen, have heard. Beginning with what they have seen, they sometimes venture to ask themselves how it came about. Because that's the way it works? They compose anatomies, with pieces of the body. Variable geometry. All living beings redesigned. Its sufficient that the organs are there, that they have a place inside an envelop of skin.

At times they are caught by a sense of suspicion. 'It is very astonishing ... but well, since he said so.'

'He.' 'They.' Many:

*Crates of Pergamon, Isogonus of Nicea, Aristaeus of Proconnesus, Agatharchides, Calliphanes, Nymphodorus, Apolonides, Phylarchus, Damon, Megasthenes, the India specialist. And, of course, Herodotus.*

They wrote their remarks down in their notebooks. Having travelled, having played host to travellers. Having walked. Ethiopia, Egypt, Lybia, India. The land of the Scythians, and beyond.

When reading their descriptions, one feels deprived.

I

The ostrich of which the lowest eyelid does not cover the eye — as is quite common with birds — might be human, says Aristotle. Moreover, this heavy animal runs but does not fly. It has got feathers, though.

Human, ostrich? He wonders. Examines.

When plucking a chicken, a hen, even a cock, someone said to Plato: 'Look, here is your man', and released the animal on the marketplace. Plato wrapped himself up in his toga and flushed with anger. It caused a great deal of fuss.

Between the birds and what is called man, both two-footed, the frontier is unclear. When there are no more boundaries, there are no more boundary-posts. That is for sure. But why?

The appealing elephant is a marine animal, it ploughs through the mud of the swamps and breaths through its lifted trunk. Like the diver with his hollow reedstem. Pearlfisher.

It is a pity it does not have webbed feet.

There is always something missing, an appendix, a compartment in the brain, or there is too much.

This gap between man and animal. It is called difference, and sometimes it is very small.

Both men and animals re affected by the forbidden. Many amongst them honour their mothers so much that they cannot stand the idea of kissing them.

The camel commits suicide when by ruse — by covering its eyes with a cloth — one forces it to mount its mother. It plunges into the void.

Where was it, in which place?

When the scorpion travels, going from north to south, its mild poison becomes lethal. The sleepiness of the dog-days arouses its malignity. Let sleeping dogs lie, the scorpion, the spider called tarantula. During holidays on the isles a lot of people, brought there by tourist-cars, become wild. Or on the ice of the Andes and in the Maya-forests.

The borderline trembles in the sun. Mirage. Where is the line of demarcation?

Pasiphae longs for the white bull, which snorts and tramples in the field. Bulls and cows cross the desert, driven by the fury of an unsatisfied desire. They haunt the crossword-puzzles with their emblematic letters, 'IO'.

And sing the icy screams of the Bacchants with ruffled hair. Eagerly tearing apart the hyenas, jackals, ewes and others.

As for the elephant, it makes so much noise around the swamps when calling its companion that people in the neighbourhood become deaf. It's easily done. However, its tongue is so small that one can hardly see it.

There is no distinction between the living. And sometimes not between the dead. The one eats the other, that's nothing new. It's proven.

There are people with only one leg, the Sciapodes; the hottest moment of the day they lie down on their back and protect themselves against the sun by the shadow of their one leg lifted above their heads. And in a place called Abarimon, beyond the man-eating Scyths, others have the soles of their feet turned backwards. On the mountain named Nulus. On the borderline.

In India one tribe only has little holes instead of nostrils and supple legs like the body of a snake.

Or others, feeding themselves exclusively with the perfume of carrots, flowers and apples from the woods; they take it with them when they travel in order not to miss the scent.

The ibisses, which also stand on one leg and build their nests on high ground, catch winged snakes in flight and devour them, thus protecting the Nile plain and its inhabitants. Now they are ordained sacred birds.

The ibisses are preferable to the Lestrygans and the Cyclops living in Sicily. Not everybody can be an ibis.

The pygmies, on the other hand, catch cranes against which they incessantly struggle. Once a year they organize a military campaign on the banks of the river, where the cranes live, and destroy their eggs.

The crocodile has a reversed jaw-bone, the Indian donkey has only one horn in the middle of its forehead, there are bulls which graze backwards, because their horns are curved forward. The dolphin sleeps on the bottom of the sea, while softly moving its fins. The Egyptian hippopotamus has a mane like a horse, is clooven-footed like the bull and has a flat nose; moreover, it has a pigtail and the voice of a horse. Spears are made from its thick skin.

But it is also true that people in Egypt do everything the other way rounded, men urinate sitting, women standing.

The deserts are the territory of the animals, be it deserts of fire or of ice. To find them you have to walk there. As far as you can go. Of course one has to walk, for days; ten days of walking across the Borysthenus to see the cannibals wearing the scalps of their victims around their necks as serviettes, or thirteen days to meet the Sauromates who eat only once every three days.

## II

At this moment, of which I am speaking, the earth is as flat as a sheet of paper. If you go to the edge, you will fall off. Galileo has yet to be born. The earth does not move, does not turn. If you go on and on, you will not find the beginning. You do not come back to the starting point. Behind the columns of Hercules the water flows into nothingness. Do not venture too far. But no one says how far, so how can you know?

Stick your hand out into space and, it will disappear, because beyond the earth there is nothing.

In this way the camel which crosses the limit disappears into the void; it does not fall off the mountain, it just went over to the other side. The limit is a peculiar thing, it is not determined once and for all. It needs to be explored. Surveyed. At your own risk.

For if the earth is flat, it must have edges, borders. Do not fall off the edge.

It is said that.
    'It':
*Crates of Pergamon, Isogonus of Nicea, Aristaeus of Proconnesus, Agatharchides, Calliphanes, Nymphodorus, Apolonides, Phylarchus, Damon, Megasthenes, the specialist of India. And of course Herodotus.*

Yes, there are borders, undoubtedly, but of location, of space. For within the boundaries of the earth, this flat leaf, living things move and mingle. The mixtures, all mixtures are possible. Just go and look.
    Hands for arms, eyes for heads, but in which order?

It seems there are only spatial distinctions.
    Thick, small, long, north (at the top of the page), south (at the bottom), east where the light originates from, west where the light disappears.

Divisions according to region, climate, soil. In the centre, the best known, and therefore probably the norm. At the edges, there are the extremes, well defined. Farther on, beyond, there is nothing.

The sky is to be found in nothingness, free of the page and its geometry.
    The sky is round. Completely elsewhere. With its own logic if one can call logic which has nothing to do with logos. Inhabited by people who undoubtedly have other manners and customs.
    It is not known who. They are called Gods. They certainly are strange creatures who poke their noses into our business according to a special logic, of their own. They think circumferentially, regeneratively.
    The circumference moves in a circle, with the heavenly bodies. Different spheres circle around the earth. They have a large sidereal period. They have an eternity in which to reappear.

126

We, on earth, have not. When it is over, it is finished. The last word is said. The page filled. Here we are sublunary. We think line, period, comma, displacement. We move on roads of earth. Of paper. We run, easily, some faster than others.

The periods are shorter, variable. They are not precisely determined. Except for the laugh which appears to man after only forty days of life. As for the rest, it depends. One single law: birth and degeneration, that is the way it is. The earth is a sheet of paper to be written on. No more, no less. It is enough. Full of signs to be decoded. A whole life is written down on it. The combinations are inexhaustible.

Strangely enough little of what is to be found in this domain is universal. An infinity of possible cases, of particular cases. With shades of expression. The interesting, the rare, the peculiar take precedence over the large divisions. There are, however, a few recurring elements like the idea of the two-footed. The ostrich, for example. The animal which resembles man so much that it approximates his essence: the bird.

The monkey, no, for having no behind and only a short tail most of the time it falls onto its forelimbs and walks on four legs. A humiliating situation. Quadruped.

Moreover, they possess a hairy belly which distinguishes them from man, but they also have lashes on both eye-lids, which distinguishes them from the other four-footed. The eye-lashes bring us back to man. Half four-footed (no bottom), half two-footed (no tail). We do not know in what category to class them. Mixed.

## III

For us, the modern (?), everything is different, it is the monkey which is close to us, we think that we owe it something, a — contested — origin which serves as reference. A somewhat quadruped ancestor.

For we think with time and its vertical descent. We have forgotten the horizontal, the arrangement of the parts of the body according to a geographical pattern. We forget anatomy for the benefit of alchemistic machinations. The genes, the genomes, the chemical combinations. We think in circles like the fictional inhabitants of the sky, gods in miniature. The classics would let us play our game. Children's games, little legs thrown into the air.

They would like to continue exploring this earth, this wonder. The earth, not the planet — an unknown word.

And without moral, without advice, without good or evil, without recipes. Only descriptions.

Plutarch had already interfered and introduced an order: 'one has to, one does not have to, we have to, we do not have to...'. He argues, in the belief that he is doing something good. Not to eat uncooked, respect the animal, allotting sense to it, reason, divide territories, limit the use in name of man. He is no longer a classic writer. He rewrites and thereby changes the terms. He changes the point of view. In reality no longer 'seeing'. Moving only within the idea. Stealing the body of the animals to make 'moral lessons' out of it. Let us be patient as a dog, as faithful as an elephant, as cunning as the donkey that lowers itself and its salt-bags when crossing the river in order to lighten the burden, or as the hedgehog foreseeing which way the wind will blow.

Not to mention the beautiful example of the faithful love of the dragon sleeping with the young girl without doing her harm.

The fable comes about by incorporating stories, some fantastic, others made credible, but anyhow without the flame of astonishment which the proven 'facts' evoked among the classics. A collection of idiocies concerning animals. It originates on the curved back of the animals, which it ruins for the benefit of human society.

## IV

To leave Plutarch and return to the classic, to their worlds. Sometimes we might try to join them, to believe in the possible journeys of the bodies, that we could be birds or flies. With machines. Or with artificial wings.

'It is said that'

'It':

*Newspapers, television, computer-channels, scholars, information, abstractions.*

It is said that the Americans, covered in hermetic suits the outside coating of which is self-adhesive, jump against a wall in order to stick to it. They play the fly.

Others mistake themselves for fishes, sleep next to the dolphins on the bottom of the ocean.

To go beyond the human and his body blocked by machines and especially go beyond consciousness, in order to be able to look elsewhere. Countries to explore.

And perhaps — after a lot of detours, persistent explorations, complicated apparatus — finally getting back something of the first wonder: the earth not so round, partially flat, made inexplorable by the waste of the endless universe, taking refuge in spaces where we might fall, flat top, something improper. We, doubting our own body, ready for all combinations, for daring mixtures, aeromechanical, with pipes, onto which strange microscopic creatures are grafted, decomposed and recomposed. In short, immoral.

Haunted by the domain of the animals, by its mysteries. And, of course, death. Which draws the whole thing with its awkward, already trembling hand.

*Translated from the Dutch by Carla Venken and Gregory Ball*

## Bibliography

Aristotle, *Part on the animals, Genesis of the animals, History of the animals.*
Plinius the Elder, *Natural History, Book VII.*
Herodotus, *The Inquiry, Books II, III, IV.*
Plutarchus, *Three Treatises for the Animals.*

# The grandeur of the horse

## Fernando Savater

*The gentlemen are always on horseback, and have brought horses to an ideal perfection; the English racer is a factitious breed. A score or two mounted gentlemen may frequently be seen running like centaurs down a hill nearly as steep as the roof of a house. Every inn-room is lined with pictures of races; telegraphs communicate, every hour, tidings of the heats from Newmarket and Ascot; and the House of Commons adjourns over the Derby Day.*

(R.W. Emerson, *English Traits*, 1856)

Now that for the people of my generation — maybe even for the whole of Europe — the winter of our discontent has arrived, it is impossible not to agree with Gloucester, who later became Richard III after a series of crimes, when he says in deep despair that he has collided with his own soul and has turned into his own enemy. The winter of a discontent that not even the pale rays of York sunshine are able to heat up. The desperate battle against one's own soul, the enmity with one's self, or with what's like one's self! Isn't it time for us, European despondents, that we recognize here the most genuine voice and the key of our despondency?

However, there is one point that I will never agree upon with Richard Gloucester. I am not willing to exchange my kingdom for a horse, however much the circumstances seem to demand that. Not because I do not believe that I actually have a kingdom that is more valuable than a horse — I am no king, I have no liking for kings, I lack the calling to the royal profession, and to that of a vassal. But if one could call our 'kingdom' the indispensable treasure of life and liberty that dignifies each human being, from the lowest to the highest ranks of society... No, not even then would I want to exchange my kingdom for a horse, because life and liberty for me have the same symbolic value as that horse. My kingdom is simple and common just like anyone's: it is already a horse that is galloping with wild manes through the tumultuous air on a June afternoon. And I wouldn't want to change that for anything.

A horse — any horse? And me, riding it? Either at a walk through valleys and plains, full of admiration for the pure beauty of the landscape, or perhaps 'in intimate contact with nature', as the snobs say. No, please, keep such nonsense away from me. I have too much respect for our troubled human condition to allow myself any eulogies of nature, whether ethical or aesthetic. I promise not to get more involved with nature than she does with me, and if I do, only out of self-defence. But let no one think that I agree with each and every one of her capricious acts!

I also have too much respect for horses to believe that they are all the same or to ride them all personally. The horse I'm referring to (the one that rules me and that I have turned into a symbol of my kingdom) is the English thorough-bred — more accurately the Anglo-Arab thoroughbred — that is raised for the races. Its jockey will never be the rounded and clumsy person that I am, but instead a light and vigorous athlete, usually of small stature (apart from some sublime exceptions), but all the same tough-nerved. The setting, then, in which the horse acquires its full glory has something of garden and casino, of park and gymnasium, without forgetting the lottery element: of course I am referring to that wonderful green and urbanized area called the hippodrome.

All true love — I do not say 'passionate' love since love without passion cannot be love, but is just a hobby or a mediocre affection — all true love, then, has as its most attractive quality that it is unjustifiable. However, this does not stop us, lovers, from living our fragile lives praising and glorifying what we love: not in order to justify the love that we feel but to justify to ourselves our very enjoyment of it. Why am I enthralled by horse races instead of, for instance, a more subtle and noble activity, a more sublime art form, a science that has more influence on the progress of man? Because. Because of anything — it doesn't matter. Montaigne revealed the ineffable secret of his friendship for La Boétie saying 'Because he was he, because I was I.' I love the horse races because they are what they are and I am who I am. Psychoanalysts can ponder the fact that my father had taken me to the racetrack ever since I was five years old, just him and me, the way other fathers take their kids to go shooting or to look for mushrooms. This should explain at least something, although if I had seriously detested the races or if I had simply taken no interest at all, that would also mean something. Sociologists can delve into my social background and geographers will no doubt remark that I grew up in the same small Basque town where I was born, close to the oldest and most elegant Spanish racetrack. Such are the circumstances, perhaps even the necessary preconditions, but they will never suffice. Beyond them all and beyond their combination, there is — enigmatic and impenetrable — the element of absolute ravishment.

Of real importance is that the experience of love, of any love, constitutes a fair emblem of what it means to be alive as a human being. All love generates and maintains a microcosm in which nothing essential is lacking. Passion reveals the essence of what we are and what we do; it provides us with a *Weltanschaung* in bonsai-shape. That is how it went for me and the horse races, where I found the basic metaphor for a culture that both prolongs and negates nature: the English thoroughbred, a living piece of art, a natural artificial being. In the elegant saying 'glorious uncertainty of the turf' we find reunited the ritual weight of tradition (in terms of the post-feudal aristocracy exhibiting the colours of its house in the jackets of its jockeys) with the demands of the new mass democracy, that is avid for big spectacles and quick money, all of this as arbitrary as sin. Here memories come back to me of the heroes and the painful downfalls of the demigods; the youngsters in whose early efforts the legends of their forefathers reappear; luck triumphing over merit; and merit as the highest and most inexplicable form of luck; the urgent necessity to prove yourself, not when you can or want to, but when you have to; the admiration for what irrefutably is, together with a latent sense of sympathy for what cannot be; the dexterity that compensates for weakness and turns it into strength; the existence of free competition among equals, showing that they never are each other's true equals; and a calendar full of unavoidable appointments that signals for me the points of reference of the yearly routine, allowing mythology to seep in with its references to the great moments of the past. That's why I talk of emblems and metaphors. The essence of existence shows itself but does not explain itself. We feel that everything is within reach, yet we continue without grasping anything. It is worse even to want to grasp everything, since this would amount to a megalomaniac joke of philosophical modesty. José Maria Alvarez phrased it quite well (especially because obscenely so) in his novel La caza del zorro: 'To want to understand the world... This useless effort always reminds me of the image of princess Maria Luisa that afternoon in the Santa Margherita stables, as she tried to suck the horse. The only thing she accomplished was to choke herself.'

Sometimes, however, the precise and beautiful quality of the metaphor (although there may be other, no less beautiful and precise metaphors) brings us closer to the partial understanding of something like a revelation. Then we are struck by rational lightning, and that is sufficient for the level-headed mind that is well aware that philosophy is not a matter of accounts and calculations, but of tales and stories. As an ethics teacher, I always struggle with the problem of how to make understandable the requirement that precedes all possible moral considerations, namely human freedom. The students always see with greater clarity the compulsory character of our condition, our limitations, and the imposition of the environment, than the open availability of choice. They usually call 'free' not the possibility to choose or to attempt to make a choice, but only the idea of succeeding, despite so many occasions in which this idea has been contradicted by the circumstances. Damned theological contamination, which confuses 'freedom' with 'omnipotence' so that we depreciate the former but appreciate the latter more! But I know what freedom is (although perhaps I can't always explain it) thanks to my frequent visits to the racetrack ...

Let's see. The jockey is free, and therefore able to do his work well or badly. He is confronted with many, sometimes decisive factors: to begin with, there is the

horse he is mounting, whose abilities do not depend on him, nor does its ath-
letic condition, given to him by the trainer; the jockey does not have a say in
the distance to be run or in the state of the track, which can be dry or muddy
depending on the meteorological circumstances; furthermore, he can't choose
his competitors and eliminate the most dangerous ones: only the capable ones
run, and only the ones that want to. The race is, from start to finish, full of
events beyond his control. Take, for instance, the starting-box that is assigned to
him by lottery. Or the pace that the competitors will establish, perhaps swift and
exhausting from the start, or falsely slow, restraining itself for a fast final run:
Will our free jockey have to save his strength for the last few metres, or will he
have to act immediately in order to wear out the ones that want to follow his
rhythm? During the run as well anything can happen: probably our man will
become surrounded in the pack, his attack obstructed by exhausted horses...
The bold one that escapes in a powerful gallop to several lengths in front of the
rest can be just a simple 'hare' that sacrifices himself for the benefit of another
and soon quits, or he can be the most dangerous of all the competitors ... Does
he have to attack now, or wait a little longer, with the risk that he's waiting too
long? Well then, poor free jockey: so many circumstances that slip away
through his fingers and that cast suspicion on him! Nevertheless, in spite of all
of these coincidences that someone who really knows how to watch a race is
clearly aware of, it is obvious that things can be good or bad, that the race can
go smoothly or be full of errors: that is to say, there is freedom. And as freedom
demands it, it doesn't matter much who wins or loses the race: sometimes the
best jockey finishes in the third place with a horse that was bound to finish last...
Good, truly good is not the one that always wins, which is an absurd (theological) pretension; nor the one that thinks he has the right to blame all his problems on the circumstances, nor the one that refuses to compete as long as
circumstances are favourable to him, or at least equally favourable to everyone,
which are two immobilising forms of the impossible. Good, truly good is he
who as a rule does not lose with the horse that is supposed to win, because he
does not mount the horse as a loser. Or so it was said — more or less — by one
of the greatest masters of horse ethics, Lester Piggott.

And that's it. I talk about glorious horses but also about everything else. If something has to be added still, it can only be what psychoanalyst and anthropologist Geza Roheim said about how one can characterize the origin and the function of human civilization: 'theΔ143
 remarkable efforts of a baby that is afraid in the dark'.

*Translated from the Spanish by Patricia de Laet*

# Cows

## *Witold Gombrowicz*

*Wednesday*

I was walking along a eucalyptus-lined avenue when a cow sauntered out from behind a tree.

I stopped and we looked each other in the eye.

Her cowness shocked my humanness to such a degree — the moment our eyes met was so intense — I stopped dead in my tracks and lost my bearings *as a man*, that is, as a member of the human species. The strange feeling that I was apparently discovering for the first time was the shame of a man come face-to-face with an animal. I allowed her to look and see me — this made us equal — and resulted in my also becoming an animal — but a strange even forbidden one, I would say. I continued my walk, but I felt uncomfortable... in nature, surrounding me on all sides, as if it were... watching me.

*Thursday*

Cows

When I pass a herd of cows, they turn their heads towards me and their eyes do not leave me until I pass. Just like at the Russoviches' in Corrientes. But then I paid attention, whereas now, after the matter of 'the cow who saw me', these looks seem like seeing to me. Grass and herbs! Trees and fields! The green nature of the world! I immerse myself in this expanse as if I were pushing off from shore and a presence consisting of a billion beings overwhelms me. O pulsating, living matter! Resplendent sunsets; today two white-and-coffee-brown islands — mountains of towers of glowing stalactites — rose before me in a crown of rubies. The islands melted together creating a bay of mystic azure so utterly without blemish that I almost believed in God — and then dark, creeping billows gathered right over the horizon — just one luminous point, a single beating heart of light, remained among the deep brown bellies of clouds crowding the horizon. Hosanna! I don't really want to write about this; after all, so many sunsets have been described in literature, and especially in ours.

I mean to say something else. The cow. How am I supposed to act toward a cow?

Nature. How am I supposed to behave toward nature?

So I head down the road, surrounded by pampa — and I feel that I am a foreigner in all of this nature, I, in my human skin… a stranger. Disturbingly different. A separate creature. And I see that Polish descriptions of nature, like all others, are worthless to me in this sudden opposition between nature and my humanity. An opposition clamoring for a resolution.

Polish descriptions of nature. So much art has been invested in them with what hopeless results. How long have we been smelling the flowers, basking in sunsets, immersing our faces in clumps of spring-green foliage, inhaling early mornings and singing hymns in honor of the Creator: who thought up these wonders? But this humble and profound prostrating of ourselves, kneeling, sniffing and smelling, has merely removed us from the most unrelenting human truth — namely, that man is not natural, he is anti-natural.

If the nation to which I belong had felt at one time that it differed in its essence from a horse, it was only because the Church lectured it about the immortal human soul. But who created that soul? God. And who created the horse? God. Thus man and horse merge in the harmony of that beginning. The contrast between them is reconcilable.

I am getting to the end of the eucalyptus-lined road. It is getting dark. The question: am I, deprived of God, closer to or farther from nature as a result of this? Answer: I am farther away. And even this opposition between me and nature becomes, without Him, impossible to mend — here there is no appeal to a higher tribunal.

But even if I were to believe in God, the Catholic view of nature would be impossible for me, in contradiction to my entire consciousness, at odds with my sensibility — and this because of the problem of pain. Catholicism has treated all of creation, except for man, with disdain. It is difficult to imagine a more Olympian indifference to 'their' pain — 'theirs', the pain of plants or animals. Man's pain has a free will and, therefore, is punishment for sins, and his future life will make just amends for the injustices of this world. But the horse? Worm? They have been forgotten. Their suffering is deprived of justice — a naked fact gaping with the absolute of despair. I am bypassing the complex dialectic of the holy doctors. I speak of the average Catholic, who, walking in the light of a justice that endows him with everything he deserves, is deaf to the immeasurable abyss of that — unjustified — suffering. Let them suffer! This does not concern him. Why, they have no souls. Let them suffer, therefore, senselessly. Yes, it would be difficult to find a teaching that concerns itself less with the world beyond man (the ahuman world); this is a doctrine proudly human, cruelly aristocratic — and how can we be surprised that it has led us into a state of blissful unconsciousness and holy innocence regarding nature, which surfaces in our idyllic descriptions of dawn or dusk.

*Translated from the Polish by Lillian Vallee*

# My bestiary in instalments

*Jacq Vogelaar*

*Song one*

If it was a quiet time and a peaceful village, or, if you like, a town (where our drama takes place) a peaceful town, of course, a non descript town, then the people in those parts must obviously have had something peaceful about them; for forgetting is a precondition for continuing to live, certainly if it was a life that was handed on from generation to generation. Forgetting? Did I say forgetting? Seeking, I meant to say, losing, thinking away, putting a full stop to the past. And did I say that I went? That can't be true. I hereby declare it to be untrue. I went, in a different sense, that is, I changed places just like someone who gets beside himself. I went into the nearest hotel, if the town had had a hotel, and sank my melancholy body into a purple almost flesh-coloured tub chair and waited, as befits an outsider, for things which were going to come and are still coming. I went and I waited. Thus far the words of someone whose name escapes me but who looked suspiciously like Mor or Tor. Oh, I forgot something else that he must have said as well: It's still raining, dear people, it's raining all the way down, keep on sleeping. That was when people thought radioactive fallout was a kind of drizzle. But come on, I've got to go. His story was nearing its climax, any minute the train could be bursting into the station — and, if we don't watch out, rush by; a lost chance. I never saw him any more, even if that doesn't mean very much in itself. And is what he had to tell any the less true for that?

We were talking about a quiet time and a peaceful village, let's call it Dobbindyke, and now, I have resolved, in revenge (I add in all honesty) that the village people just had to face up to it. Real horse-lovers they were, as I was saying, even then, still are, these days they still do everything with horses, just as in the village of Pancras where they used to do everything with pigs. Dobbindyke ran on horsepower, they wanted nothing to do with electricity. They used to say there: water and electricity, it sounds like water and head or straw and man or bird and scarecrow, it won't hurt you, but put your fingers in the socket and you'll sing a different tune. Very pithy, very typical just like their whole story gives the horse away.

Now the facts. Oh yes, the facts, and the man behind the village galloping along the reclaimed Zuiderzee. Timid is the word, shy is also possible, withdrawn, but once they were galloping you wouldn't recognize them, so noble, so russet, so out and out sneaky. No, they didn't eat horsemeat, horsemeat was mushy from the whip and the curry-comb they scorned, it tasted squashed, and what your arse sits on you certainly don't eat! On the other hand, the eyes were a real delicacy for them. They stuck pointed straws into the melancholy horse's eyes and sucked them dry. What was left over from a horse was for the flies, for the flies Dobbindyke was a veritable paradise. The climate, too, played a role, sodden, and the local dialect was no less boggy, what was elsewhere called consumption was there a drizzle of words. But that must be rumour, I've been told, gossip, since from my own observations I can certify that all that talk about delicious horse's eyes is not true, on the contrary, the horses without any exception wear blinkers. The fairground horses provide the proof, they are still alive and happy, as the music which they love verifies.

Yes, the reverse is indeed the case, they loved their horses like themselves, they bent over backwards for them. No, whinnying wasn't forbidden either, only the whinnying of stallions, or at least of geldings, and then only on Sundays, since on Sundays, yet another proof of their love of animals, you could see the villagers riding through the village and the surrounding fields, proud I said, the image of their mounts, proud as centaurs but uncoupled, and so freer, more human too, nine or ten villagers on one horse, never less than eight otherwise the nag would stray too far from home, perhaps into the Dobbin or certainly to the house of Doctor Fox or to Bald Hillor or even Pancras. The villagers weren't such high-fliers. But no, that too seems a downright lie, there was always happy whinnying — so happy and full of spirits that a new word will have to be found which expresses that oneness of spirit, herd instinct and popular joy and at the same time allows the dissonances, undertones and pregnant pauses to come out: a word somewhere between burning and bleating, roaring with laughter and stammering, but let's leave that aside — and the horses, this needs to be added for the sake of the truth, ate only smaller animals of which there always seem to be enough in nature.

The villagers at that time were peaceful citizens like all the others, the fear of God was still deep in them — we're talking about 1974 — so that they were content with little and didn't do very much. All the people in those parts used to eat enormous quantities of horsemeat, how could it be otherwise, day and night they used to stuff themselves full of great hunks of horsemeat; and if they happened to be doing something else they talked about nothing else than the great servings of horsemeat they would shortly devour, horsemeat that was recommended in all the baby-books as *the* means for strengthening mothers emaciated from tuberculosis. One or two people used to speak about the side-effects of eating the noble parts of the noble animal, but most people weren't so fussy. And every hour of the day horses used to trot around and you could hear the sounds of their hooves. Pigs were taboo, as was marriage to someone from outside the village; girls were marriageable as soon as they could ride or be ridden; and for getting stuck into a rump of horse that had not been fried, roasted or boiled, there was severe corporal punishment administered by the horse in question — many an unruly village boy bore the mark of a horseshoe on his face and from time to time had to take a battering. So much for a brief introduction to the local customs and rites of initiation.

*Song two*
But I'm getting cold; the people here live in stables, which I'm not used to. I'll tell you what I've been told. Someone takes a horse, cuts a hole in his chest and

through the hole takes everything out from inside the horse, not only his speed, his fire, his universally valued trustworthiness and oft-praised readiness to help as well as his noble nature, but also the intestines, the flesh and the bones, so that in the end there is nothing left over except an empty skin, merely a casing but contrary to all logic still the packaging of a 'horse'. Perhaps that explains the other story about the animal haters who apparently ride around wildly on one horse, ten at a time. It's more likely that someone's imagination got carried away, that happens. We were talking about these emptied out horse skins which are filled up. For a while the rumour was going round that grooms had been stuffed in the horse skins on which the farmers rode, as the saying goes, regally around, as if they were gentlemen farmers and their horses farm workers in disguise; it could have happened, but it seems that it's not true. And again, it was a time when the word peace still had some meaning, even if it was beginning to sweat and ferment, when sausage was still real sausage and not some vomit mixed up with lard and sawdust in a horse's condom. No, it's like this, then I have to be off. The emptied out horse skin was filled with another horse, mostly a somewhat younger horse or a horse from a smaller race, and only if there was no horse available was it replaced by something else, if it is an animal, a him if it was a stallion, a her if it was a mare and an it if it was a mule. By the way, it was a time when both bike racing and trotting were still clean, he said significantly, if you take my meaning, have a look at my eyes, and he turned around, that is he turned his magnificent sinewy body with its plaited tail and flowing mane and gleaming rump (once and for all) around.

## Song three

For what you have seen you don't need to look any more. Mor is the author of this truism. Imagine, he says, the way he says imagine for practically every sentence, as an overture, imagine that you could recover images on a retina, from the millions of images that have ever been projected onto it, as if you are flipping through millions of paper thin images stacked on top of each other and chance — I'll speak about chance in a moment — upon that one comic strip: picture puzzle or rebus.

It was because you weren't looking for it that you found it.

Mor spoke about chance, stroke of luck to be exact: that one picture that pops up and covers up many others — which becomes *covers* in his bastardized language — others, then, as if it were a synopsis of them.

Breathe out, he says and doesn't even give me the chance to register the stage direction for his and my performance, let alone give a true-to-life description of his appearance, his physique, physiognomy and relevant character traits — malicious, always furious; but outwardly calmness itself, apart from the sarcasm that frothily drops from the corners of his mouth — what I want to say is: every time I put that famous old Mor on stage, he begins by distancing himself from me and if that doesn't work, because I don't let myself be pushed out of the way, which he knows, because he wouldn't survive it in one piece — and what, I ask you as an aside, what if he survives it and shedding his skin, as vulnerable as a wind egg, starts ranting and raving somewhere else, I ask you — he shuts me up, because he absolutely can't stand asides, or worse, he speaks on my behalf, puts his words in my mouth, and imperturbably operates on the slogan: *If you stand up you lose your seat.*

*Song four*
When I see it the way — thanks to Carpaccio — we were able to see it four centuries ago, horse and cuirass polished up as with stove polish.

From the right a permed knight charges the dragon, who with spread wings and tail curled in rage, with nothing else than natural weapons, will bite the dust. On the one side man and horse on its hind legs and on the other side the monster on the two hind legs of a lion, they form an arch through the lance which sticks in the mouth — and breaks.

Under that bridge of violence the dead lie spread-eagled: he who suffered in his flesh, Carpathius, his right leg ripped off at the groin just like his sex; and she, slumbering just as peacefully, at least what there is left of her, the upper part of her body intact to where the lower part is completely ripped off. Gigantic jaws with teeth like a fox trap have ripped off the male and female parts. Their remains are lying among black snakes, scorpions, closed crustaceans and skulls.

The poet, because naturally every knight turns in due course into a young poet and vice-versa, he would begin his story as follows: Who ate the toad's

*Vittore Carpaccio,* St George and the Dragon *(I), c. 1507 (Scuola Degli Schiavoni, Venice)*

warts? Who licked the syphilitic's dick? Who stuck his face in the pussy of the pregnant slut? Who got sores on his lips and a rash round his mouth and fungus on his tongue from nonsense and infected snoring?

Yes, we're in the margin of the fourth song!

You'd better begin, I interrupt, by quartering the picture, then the two bottom parts will make up the parties to the duel, amid all the vermin and rubbish, and the top left area consists of the geometrically ordered town and the top right section is taken up by the blue water where two sailing ships will disappear, unseen by the Stella Maris, the maid in armour.

What is there to stop the seer from saying that this emblem contains our whole culture, including its prehistory? Then it is no longer a picture but an illusion, a succession of solutions flashing past our eyes treacherously fast, so fast that to the naked eye it is standing still. That's the riddle of the picture. Deception, I think, it can't be anything else. And what else is there to do about it than think it away? Act as if attacker (defender of everything which exists there as imaginable: the whole culture, beginning with the virginity of the living and dead flesh) and prey (the enemy of order, flesh that devours itself) are stray silhouettes, shadows from another theatre, memory for instance.

Let's magnify the details.

Yes, the living are always right.

I hear Mor, the gentle healer with a stinking gob, chewing the fat after his meal of humans. It must have happened in some previous time. There remain only silent witnesses. The sex has disappeared, God knows for how long, and who knows if the male member has turned into a lizard or a toad and the female parts taken on the appearance of a crustacean, silent in every language. And don't forget the foot, lightly supporting the head rising out of the ground like a mushroom — a living skull sniffing but preferring to keep its eyes shut, the skull of wisdom — 'I know the foot', I know the riddles about feet. Run away before evening falls and colours the evening land blood-red from shame whenever this age-old spectacle unfolds, always against another backdrop.

We look and what we have seen afterwards is a metamorphosis: time blown up into space.

The dragon has his place in the genealogy of gods and people, as the fruit of violence (Ares) and eros (Aphrodite).

For eight years Cadmus had to serve Ares as a slave, then Zeus gave him Harmony as a wife. And whose daughter was Harmony? Right, Ares and Aphrodite's, from whose union the ideal woman and the dreaded monster were born.

You can see a pattern in the creation myths: there is a spring guarded by a dragon or a snake; the monster is vanquished by the hero who establishes there a town or community in the same place, or in place of it (stand up and you've lost your seat).

Whenever dragons guard springs or have them in their power, like the dragon Cadmus slew before he founded the city of Thebes, or live in swamps, like Siegfried's first dragon, have a woman in their power or guard her, like Andromeda who was freed by Perseus, or threaten to devour her, like the maiden freed by St George, or else, as in many fairy-tales and myths, guard a treasure, like the dragon and the Golden Fleece or Fafner and the Treasure of the Niebelungen — the battle between the hero and the dragon always represents the same relationship to nature.

The subjection of nature — in the language of myth always presented as a battle with a monster — is an act which lies at the foundations of community life, but it must also always be repeated, and if that fails, the result is sickness, decline and finally the death of the community.

The partial success or failure of the hero expresses a state of nature which is ultimately self-destructive for the society, a trial by fire which in the long run it is not equal to. As long as subservient and suppressed nature continues to remain suppressed, society is threatened by the uprising of the subservient spirits which it has exorcized and exploited.

*If I cannot bend the higher powers, I shall set the dark subterranean forces in motion.*

Mor knows better: Cadmus wanted to fill the serpent full of young. His lance broke out of sheer desire.

Everything achieved, smirked the strange horse doctor, from the willing womb of violence.

146

## Song five

Oh come in my mouth, gruesome beloved, open me and I shall open up, my lips licking flames, my teeth bared and lined up like rows of flashing weapons; I open up wider than I ever have (bigger words have never left me) and you — jerk your lance deeper than you thought, into my deepest secrets, where, though you did not know, someone other than me is settled, dark like no other, an adders' nest of never-sleeping thoughts brooding in my stomach; and if you now force your way in through the open door of my mouth or the open door of my burning womb, I take you while you take me and I let myself be torn apart by the barbs of your weapon.

That's what you dreamed about — with eyes open, that's why I could see it: how my annihilation would cost you your life, because, once lost in the labyrinth of my interior, you lose all understanding, first that of time, of before and after.

## Song six

The spectators have done it, not for the last time.

In front of the elegant palace stands an oriental castle where they stand and watch on the towers, battlements, galleries and walls. They will survive it, the just, they are keeping watch so that order and peace be maintained. The city of Cadmus has become the city of the Law. At a safe distance the townspeople enjoy their view.

The dragon has long been drunkenly wild, watched full of desire from behind and on the walls, from the protection of the high-built town. They followed the monster, in all his movements, like a suspicious figure and, sooner or later, suspicion always finds its confirmation. What were they waiting for? The only question was who would begin. They preferred the alien to provide the occasion by attacking an innocent victim. That would be the permit to start attacking mercilessly. You could also provoke it.

And didn't the couple themselves bring it on themselves? They became bolder and bolder; there could be no blessing on their love, breaking the law could not go unpunished. It seemed as if no one noticed when those two

147

slipped out the town gates so that, far from all reproving and monitoring eyes, they could give themselves to each other in the remains of paradise, having it off with each other, in the words of the chronicler.

The young poet — this was when the cliché was coined: the romantic revolutionary alias the ladies' hero alias the poet alias the black sheep — had said to his beloved: 'Paradise is under lock and key and a warmongering angel is keeping watch; let's make a trip round the world and see if there is a back door to paradise.' A short trip can also be a world trip and, rose-tinted as their vision was, they saw in the bedraggled little tree a descendant of the tree of knowledge.

Every step, every danger of theirs could easily be followed from the watch tower, the more so since they were so lost in each other they no longer saw any danger. Under the bare trunks of the palm trees there was no undergrowth, so that they were visible everywhere. Moreover, to make the view unhindered, even the only tree had been cut down to eye level.

And what had been hoped for took place — as if there were a scenario, it ran exactly according to the book — their lower body was torn off by the Beast. People had been waiting for this. The punishment was at once an example — of the voraciousness of the wilderness, the annihilating drive of the beast, hungering after bodies, preferably the body of a maiden. An end had to be made to unrestrained desire, away with the source of disorder. Only a shadow could be allowed to remain of the beast, an heraldic figure as a reminder of and talisman against danger.

They voluptuously looked at how the lust down below was wiped out. The spectacle was so engrossing that the men who were looking over the shoulders of their ladies took the women from behind while standing, Mor related, and he knew what he was talking about. He also talked about the return of what was suppressed and once he was away he talked further about combining words and words covering each other, and finally about an unequal struggle between two sentences.

But, at the *moment suprême* the lance broke after he had cleaved through jaw and skull and come out the other side of the head.

*Vittore Carpaccio,* St George and the Dragon *(II), c. 1507 (Scuola Degli Schiavoni, Venice)*

Adversary and plague have formed a bow for centuries, yes and no keep each other in balance: without the dragon Saint George would not be a hero, without the knight the dragon would be no threat marked out for death. George Goodblood founded the firm Rent-o-Kill — said Mor, charlatan among the crawling and flying vermin.

*Song seven*
Let's talk about the other sun for a moment, the one which is temporarily away, invisible. As long as you can't see it you might think it is negligible and insignificant, something like a blind spot, but be careful, it is in fact, virtually a system so extended that we can be fully absorbed into it. And it is Mor who is speaking, his second voice, the first person plural betrays him. If we could get her right out of those intestines, like a snail out of its shell, we'd be able to talk to her, velvet soft Pupilla, softened horse's eye, show yourself, we won't hurt you, for us you don't smell, certainly not, come on, release me, fantastic mother whom I've called Nora. She doesn't have any sections or divisions, she doesn't cut and bite. Neither does she allow herself, under whatever pressure, to be detached from the inside of this chapel which serves as her fortress, an imitation of course just like everything here is an imitation, even this day is an imitation of yesterday which under the protection of darkness is not to be distinguished from the other days. She becomes deaf and blind, she doesn't betray her feelings. She is the discoverer of panoramas, landscapes, cityscapes, vistas and flesh. She loves rumours, scandal-mongering and long-windedness, vague noises, fuzzy concepts, confused conversations, expansive gestures, unnecessary actions, feints in which she can give herself completely. Mud is her element and because she is nothing but an intensely black pool, a pitch-black mirror, a porous surface in which the rain is completely absorbed, she has, just like the rain, an aversion to discriminating distinctions. The sun which is thought to be in the firmament dissolves and dissects, it literally turns into stone everything that moves or appears to move. The soft sun, the absent one, guides. Not for nothing does mating take place in the dark — two words mount each other and ride each other together — the coupling is black. Objects also mate. What I am is one thing, how I feel is another thing, what I think of myself and how I see myself

is yet another thing, we mate in the dark, in all quietness, we purify ourselves in the mud and our thoughts burst from against the surface like bubbles.

But the fathers said: young fellow, leave your mother and love us, for we too have something that looks just like a vagina. And in that context Carnac gave an ethnographic example. Among the Poro, he said, a young man was eaten by a crocodile and lived for a time in its stomach. (Crocodile, alligator, dragon, snake, pig, they're all disguises!) We put the novices somewhere for a short time where it's nice and dark. They come out completely disorientated, often they won't come out of themselves. When they go home they have forgotten how to walk, they're unsteady on their feet and they go in the back door of their parent's house, don't laugh, and when they get their food for the first time, they hold their plate upside down. They have to learn everything again, even how to talk; the only thing that sometimes gets them talking is blowing smoke up their arse, a good smoking out is a good help.

But don't be distracted, I say, we were talking about mating, maybe I meant a mating, I find no other word fit here, the one who in a dark corner of the room, invisible to me, reproduced himself, here too I can't think of a better word so I'll say: Echo's ego, since it is perfectly possible that one person, the rightful owner of my voice or his representative, Mor or an animal vicar on earth, that this person with impure intentions had bent over the other, me as indirect object, and that the other with equally impure albeit other intentions had it off with the one or even fertilized the other, me as accusative, all of which is extremely complicated, you have or are a complex, according to the wonder doctor. I couldn't follow the rest of the scene any more, even if I wanted to.

I say: I see. He didn't have to show me, let alone do it for me. I see in the window a circular network of concentric rays and circles. A great female spider with a provocative cross on her hairy rear is sitting in the middle of that mysterious thing of hers, she's praying. I hear my mother reply to Lips's question, how do you catch your prey: I increase my hunting potential, I sharpen my watchfulness, I broaden the possibilities and the reach of my eyes and legs, I spin a web of gossamer and sensitive detectors, from a central position I am in direct contact with the farthest corners of my field. That was her lesson for life, she used her passive capacities as aggressive, possessive means. The female or my

*Vittore Carpaccio,* St George and the Dragon *(III), c. 1507 (Scuola Degli Schiavoni, Venice)*

mother in the metaphorical sense sits and watches without moving, eight legs spread ready until the smaller, probably male animal, smoother of skin, walks right up to his, or should I say her, goal and embraces her intimately. The union, or should we say reunion, is over in a flash, the worn out male body lets go of the female body that, now the procreator with madly thumping heart is hanging exhausted beside her, looks even bigger. The female seems to recover, the legs curl up and the love contraction disappears, a hairy tongue comes out,the feelers on the head move back and forth. And she uses that unguarded moment when my father is recovering from his pleasure to, lightning fast, (in actual fact, it was a question of decades) stick her claws into the bag-shaped underbody of the poor male who, after a few thrashing movements and a ridiculous attempt to escape from her iron grip, stiffened and disappeared excruciatingly slowly, to my mind, in the thoughtfully masticating mouth of the spider. Poor mother, he said, go away from that window.

*Song eight*

Well, and George marched out to fight the dragon and to free geometrical order from the danger of a confusion of lines which is always lurking around every corner, paraplegia of perspective and ordering overview of distance — from a relapse into chaos without contours and boundaries. But as calmly as he had drawn up the plan and coolly and hygienically carried it out in his head — if it had to be, then quickly and silently, avoiding any *unnecessary bloodshed* — so violently did his guts protest; a dark premonition worked its way up to his mind, a smouldering awareness in his guts that he was entering the ring with something that he knew well, all too well, indeed from too close.

He attacked the monster as if out there something manifested itself which belonged inside him, no, did not belong inside him but as a parasite had regally taken up residence in his intestines.

That hair's breadth between intending and foreboding or between knowing and seeing became a hesitation, a wavering which could have broken itself over him and did, in fact, upset him, even if this was not immediately noticeable and the consequences did not make them themselves apparent for a time — stay of execution as the prompter calls it.

If I had to write about that history, said Mor, the sardonic spell-breaker, then I would expatiate on the possibility of a role switch. Don't you think that the male victim is the brother of the knight, with all that that implies given the saint in the ground who has been cut in half, and the satanic virgin in the side aisle, the better half and the best actress. What a pity they're too big.

Why not see the attacker as a centaur, Mor went on pontificating, what happened to this self-exalting person as a result of always riding himself?! From the holy city of women from ages past we plop right down in the middle of Dobbindyke in the year 'now', key word *mud*. And don't hesitate, he blathered on, to attribute to the dragon a Gorgon's femininity. Besides, flute player and horror mask are a good match.

Tongue against spear it is, back and forward speechlessness which roars and flames — yes becomes half no, words turn into mud again in the mouth, where language comes to lie fallow like earth before and after, the tongue like an amphibian hidden away in the wet clay. But those are pictures from an in-between-time, she forgets.

The centaur storms forward — for us sideways — with his destructive organ extended, driven by the frenzy of the expression.

Mor went from bad to worse, from tongue language to linguistic nominations. Because real nature has been lost, he quoted Pascal or Lezama Lima, anything can become nature; but, he spoke prophetically — the watcher of cartoons or videos — the horrible interval can only be filled by the image, and so unceasingly he himself filled up with small talk and pictures what he had vaguely seen and heard.

We have seen it with our own eyes, but what it meant we only know by hearsay (try giving Echo a tongue kiss).

*Song nine*
Imagine the man in armour on his horse rather as an insect, the knight changed into a black scarab locked in combat with a scorpion. After that the relations are perhaps restored a little. If not willingly then unwillingly, as in every assimilation.

The spear incited lust, time after time the womb seemed to be as fertile as a volcano.

Holy dung beetle and scorpion of self-destruction, drawn or seen according to the right scale, are between them no more than a detail among details — but, no sense without details — a convulsion amidst a self-generating war between culture and nature, on a battlefield amidst mountains of skulls, jawbones of donkeys and corpses, rib cages and separate limbs, crawling vermin — out of paradise on all fours or flat on your stomach or, better even, like the animal with two backs stuck together by desire, rolling and mating at the same time, by the gate at the feet of an archangel ejaculating and coming with exultation — among lizards, snakes, frogs, which twisting their tongue, speak the language of Adam, Ammonites and other linguistic and crustacean animals, not forgetting the necessary symbols, the oyster as mother-of-pearl cosmos, in memory of the womb which without flinching saw its mollusc sucked empty and, with the whole underbody, devoured.

If only we could see what goes on in the stomach of the hydra, Mor suggested and gestured as if he, armed with a crowbar, brute force and a calf's rope, wanted to get a foal, but got himself woefully tangled in thought coils around Parsifae and her oestrous cream machine (the automatic Venus, as we can see in the sphinx-imitating Cow).

Hear the echo of the conch! reverberation of a glottal story. And sooner or later you will get lost in your own coils or otherwise in the jumble of the word and the chtonian dark — as long as things are given a name, spoke Mor threateningly, like a God swaying with his engraving knife. For my part, make things up freely, mate, turn it into an inner struggle between the hybrid centaur and the aardvaark armed with sheet metal and long teeth, and let Circe, as voluptuous as she is ascetic, look on.

It happened to her, in time and place. Isis, let's say Isis, is looking desperately, before it gets dark, for the remains of her husband lying all over the field and pieces him together as a scarecrow. But a small piece is missing, she says, a gem. That's in his tin says Alma and points to the stranger among us — but in Dobbindyke it doesn't take much to be a strange person with the accent on strange — the tin must and shall open and yum says the dog without name, the bird has flown, *the dicky bird has flown*, wails the newly wed heavenly bride.

He has betrayed himself now he has broken the rules of the house, the ver-

155

min (fill in). Before he came peace and quiet reigned there, it is still rural judging by the bleating of goats and the lowing of cows and the hum of all sorts of farm machinery, but harmony seems to be well and truly lost, relations — between Dobbindyke and Pancras — have been upset for good, we've lost the thread, the thread with which the maiden kept her half brother under control (half brother and half bull, put together also a sort of dragon) — of this thread only a trail of slime is left over. She had come running into the palace hall with a green dragon on a lead, when at once a knight rode up to the beast and speared it in its left eye; don't do it, cried the girl, but the beast was already bleeding from its mouth, thunderclouds gathered ominously over the head of the young knight, you're hurting my little brother she cried out. Shall we throw it away, some suggest, it's still about the gem in the case, but then in earlier times, or shall we eat it up others cry, don't throw anything away, interposed someone of the same sex, nothing is thrown away here as long as we have a war to remember, it is obvious from her that in this story nothing is thrown away.

*Song ten*

Bifrons tries it once again, now as a pig. We are sitting fastened to our seats and awaiting the return of our guest, says Mor in the living present tense, nothing betrays his presence, I was ashamed. In the past I had seen him proudly walking erect, oh no that was after, years after the famous incident. Humiliatingly, his head held high. On all fours he looked more in his element, then he still smelt what was more to his liking. Now that he is walking erect, conscious choice or not, he is nauseated by every smell that has its origin between hind legs, having forgotten that his organ of smell was once strongly directed there, whether shyly or not. He's grunting I said, not being able to disguise a hint of contempt. He's thinking was the retort. Thinking in such a subservient position, I can't accept that. He noses around. He sniffs. I say I call that snuffling, I call that rooting around, for the etymologist in me, says Carnac, there is a suspiciously strong relationship between snuffling and the Greek word *noos*, yes, ask me another one, laughs Uncle Lambert laughs when he sees the following course brought in. All at once everyone is there. It is brought in on big plates, in a word regally says Pandora, like the Pope at Easter. That language is better understood here, says Bono, welcome among us. Brothers and sisters, tries Imoff once more. Why do males like riding so much, Pandora chatters on, not without reason, she often sacrifices herself. Lama is of the devil, why do you call her like that, I ask Goose, do you still have to ask that, she reminds the guests of the proverb: A woman and a pig under one roof spoils the broth. That's true agrees someone, not just anyone, just in time they do what they say, and right in front of the door to the bride's sty the head of the cock is chopped off. Suddenly the food doesn't taste good any more, in a manner of speaking we're fed up to the back teeth, a pigsty I hear someone say on their way out, someone wants to be funny and barks, another whinnies, someone else crows a couple of times, only the grunting sounds thoroughly contented and sincere.

It looks as if O's stomach is full, drowsily he faces an uncertain future, or he'd probably call it a wise look and himself, if it is suitable, a preacher. His hands crossed at the back of his neck, behold the biggest eater, or as he would also like to be called, the master-eater, mad as he is about titles and marks of honour, he lets it be known that he has had a gutful of eating. There is something wrong with his digestion, he is continually hungry but at the very thought of eating he has to retch, and if he does eat something, preferably something sweet, creamed chestnuts with fresh cream, chocolate mousse, stewed figs with cinnamon, he can't get rid of it, which is all too obvious from the way he looks. Now and again he sticks his finger in a plate or pot of something, or picks something from someone else's plate. Tuck in boys, he encourages with the yawn of a lion. Everyone is full, most of the faces are blissful, in short measured sentences recipes are exchanged. But there's something wrong which can only be noticed by an observant listener, a ripple you might say, as it were a crease in the sentence or if you wish an interval, a suggestion of wind (the conceited pig that bites the hand of the one who strokes or feeds it, ouch, cries Antony in an unnecessarily loud voice; the pig sits on its rear in the devout posture of a dog, he too has his special wishes, I hear the voice of Janus, that must be a mistake: I'm going to look for a tree with a hard trunk, by biting or gnawing it my teeth will grow, I want to have the fangs of a wild boar, long and sharp; I want to run over the dry leaves in the woods, on my way eating up sleeping snakes and birds who have fallen out of their nest and armies of hares in their holes; I root around

*Vittore Carpaccio*, St George and the Dragon *(IV)*, c. *1507 (Scuola Degli Schiavoni, Venice)*

ploughed fields, I trample the green wheat into the mud, and I squash fruit figs, melons and cucumbers flat; and I shall cross the waters, go ashore and in the sand I shall break the shells of the great turtle eggs so that the yolk runs out; I will scare everyone in the towns to death, devour the children in front of the door, burst into the houses and dance on the tables and upturn plates and bowls; by scratching on the walls and digging under them I shall tear down churches and big buildings, I shall root around in the graves to eat up the rotten princes in their coffins and their molten flesh shall dribble over my lips and chin; I shall grow, I shall swell up, I shall hear all sorts of things rumble in my stomach and make unmistakable noises — 'Why are you biting me?' asks Antony in an aggrieved voice, 'Why are you so angry?' — what do you think, that I can live on those bits of vegetable leaves and leftovers you give me? and why have you treated me differently, daddy, you picked me out from among my brothers and tied me by my ears to your belt and brought me here, my mother cried, I cried the wet lungs out of my young body, but you went away without the least concern, your work supposedly came first, your work which you called your vocation, and still do, and still you take no notice of me, but mark my words, I'm fed up with stuffing all those letters into me, I want some juicy females, I want a trough full of golden white meal with pink froth of blood in it, I want scarlet straw and under my feet I want to hear human bones breaking like dried grape vines; and beginning with you, old sucker, you great bore, I'm going to bore a hole in your side and drink your gall — you ungrateful pig, squealed Antony sucking his painful finger). Sitting around a table is a conciliatory act, even if it is a last supper, that is why we are together, there is something to be made up for, oh sum up this happy gathering even if superfluously, let us for a while forget all our differences and disputes, Imoff repeats it more elaborately once more, Carnac replaces some words with others, less direct expressions more amenable to all kinds of explanation, credible expressions, so that no one has the feeling that he has singed his wings or has had his pride hurt, regardless of whether it is the pride of a lowly person or the greater pride of a superior person, pride is always unreasonable and is above all a question of words like shame, happiness and unhappiness, and being absolutely right, somewhere in a corner a discussion of strategy is taking place. Now, Mor whispers to me, grab your chance. Ask where the cemetery for pigs is or go and have a look yourself for intestines and stomachs, find it in the shit canals as a result of the great murderous and gluttonous distraction.

## Song eleven

Last night I was thinking about you, you sows and boars, restless philosophers with short muddy legs which you drill in the ground like alpenstocks; your stolid bodies, bloated, faithless frogs; how must I disguise myself to outwit you and discover how, rooting around, grunting, trampling everything to slush, you snuffle out of muck and stupor the hazy pearl of a wisdom that is as absurd as it is unfindable, the fruit of a lost grain, a body gone astray. Why was I thinking about you last night? Because something snapped in my breast.

Why, I thought last night, am I akin to them?

I sang out, I thought, but Mor, always wanting to have the last word, added another footnote to my swansong. Not only did he explain to me in more detail with the bold assertion that (unconsciously) I had to recognize my kinship and had to admit that my friendship for birds was based on wishful thinking, on a so-called projection, and without any difficulty he cut and shaved my soul with his therapeutic clippers so that it stood in the window like a marzipan pig, for everybody to look at. In an expansive manner he explained it in stronger language, supposedly speaking on my behalf: My night with the pig took place on a clear day, in full view of everybody, though it was — to cloak the deceit by a one-sided illusion — in the middle of the day for her and me in our cage — in her pen, the piglet's, or to be more precise in someone else's cage — a nocturne being enacted: in a cage made of black glass that from the outside was transparent and on the inside seemed to enclose us like the night, and I stood next to her and ran my knuckles along her flank, where the rough hairs changed into the pink skin of her full udders, and she liked that a lot judging by the satisfied grunting and with a deep sigh she sank down and lay on her side, her hind legs twitching with desire, ready to receive me in a human, all too human way, in her wisdom thinking about what the fruit of our union could mean for the future of man and beast — according to Mor on my behalf, as regards usage taking an advance on the stark sameness and the sameness of language.

*As an encore another advance, a foretaste.*
After they had kept it (the creature) locked up for years — to protect it from itself, ran the argument, to prevent it from wreaking havoc, was the supplementary argument in the event that the first should turn out to be untenable, what they meant was: out of fear that it would cause evil, that's why they said: to teach it to make the sounds and gestures which everyone understood so that they could deal with each other as equals — after they had kept it locked up, for the first years violently under lock and key violence, after that voluntarily, they cried, when they once saw it (in the wild, for real, in this closed community) running around and heard it making noises which they were not accustomed to hear from it (the creature), which in fact seemed suspiciously like the sounds they themselves made, they cried that it had finally become normal, and contemptuously they turned up their noses now that it no longer was a bone of contention to them, so that in the course of time they developed a certain affection for it. They called it (the creature) a milksop, whereby it had for the first time got a name and actually something of a (recognizable or at least a familiar) face. Fortunately for them they had not realized that it had only imitated them as a joke, even if it was only because it finally wanted to know how it felt to be like them.

*Translated from the Dutch by Kristin Davidse*

# pig near Cassel
## tableau

### *Luuk Gruwez*

in love I like the larger sizes.
I grunt enamoured of trough and sow,
but the farmer is my master
and my master kips
yonder under the apple tree

and Cassel kips among cows and turkeys.
it's a Sunday in the month of May,
it's an aeon you have to yawn at.
I become ponderous of corpulence.

a cow can't know about a pig
as I know about the cow,
'cause I'm smart, I can cry,
sometimes I blubber through my muddy bum.
There sticks my tail, my only question.

why's my gaze so overcast, d'you know?
because I cannot help that I,
with so much of a baby's pink,
am born already old.

in love I like the larger sizes.
I stand upright in the universe,
endowed with all my chops.
I gasp a bit dreamily
at this whole mighty firmament
but fall asleep in muck and dung
— just like a swine, and just like me.

From Luuk Gruwez, *Dikke Mensen* [Fat people]

*Translatied from the Dutch by Frank Despriet and Tony Solomone*

*From Staf De Roo & Erik De Smet,* Her Eeklo van toen *(private collection: Brugge, 1985)*

# The man with the swine's head

## Paul van Ostaijen

And yet his parents were honest folk. That was beyond question. His father was a lumberjack. A perfectly honest man who, when his beloved had told him in tears that something wasn't right, that is, all too right, had immediately taken her to the town hall and the church. No matter how hard life was, they didn't grumble. Much rather, they looked each coming dawn hopefully in the face. Then the illusion shattered. When he was born they shed bitter tears. No, this ordeal was too heavy. The mother fainted onto her bed of woe and the father broke his clay pipe. 'That was all we needed', he said, threw his broken pipe into a corner and shook his fist at heaven. The boy had a swine's head without any notable effects. The father said in his rage: 'I have never seen a swine that had such a perfect swine's head as my descendant. He beats the record in swine's heads.' The father did not comprehend the full bearing of these words spoken in anger. But later he frequently recalled these wise words and began to believe in his prophecy.

After they had shed these bitter tears, they wanted to console each other. It didn't work. It led to sharp words. 'Who's got a swine's head in your family', he asked. 'Not in my family, it's in your family we should look for the cause', the indignant Christina replied. 'Or you were continuously thinking of a swine's head while you were pregnant', he went on, 'I ask myself what I have to expect

*René Magritte,* La Bonne Fortune *(private collection)*

next from a woman who continually thinks of a swine's head, and that in the highly responsible period of her pregnancy to boot. How deep have you sunk in the mud of vice to always have such an indecent image in your head? This is not an ordeal, more likely a punishment.'

The distant neighbours tried calming the unhappy parents down. The father had already spent all his savings on the replenishment of his stock of pipes. For every excitation cost a pipe. The neighbours said: 'Give it a moment's thought. That swine's head is only like that at birth. Later everything will be all right.

And you'll see what a beautiful head Anatool will have. You'll be proud of it later.' An old woman illustrated this: 'Many years ago I was present at the birth of a child that carried a church on its head. It was in all the papers. Well, that church gradually became a chapel, then a smaller chapel, then a pub, and finally a beautiful head of blond curls. It'll be exactly the same with Anatool. And don't forget: such children are usually very smart and become their parents' pride and joy.'

Later the doctor from the castle came. He pulled a very puzzled face and said: 'My science is baffled. I don't know what the child should eat. If it is a swine, as his head seems to indicate, then it must eat buttermilk with acorns and potatoes. But if it is a human being, as some details seem to suggest, then I must prescribe something very different. Truly, I don't know which way to turn the matter.' The countess visited the unfortunate parents and was full of enthusiasm when she saw little Anatool. 'Il est unique', she said, and wanted to buy Anatool, for she had a world-famous collection of odd swine. Anatool would surely be the crowning item. She promised the parents that he would be raised in a marble sty and that he should want nothing: in his sty he would be able to live a life of peace and quiet, for the countess's pigs were kept only for the sake of curiousness and by no means with the intention of being slaughtered. To the lumberjack and his wife this seemed a very nice proposal as well as a distinguished future for their descendant. It was only when the countess was so undiplomatic as to insist that from the moment little Anatool would come into her possession he would cease to be a human being and would thenceforth belong to the animal species of swine, that Christina's mother's heart revolted. As did her sense of honour. It was emphatically a human being she had given birth to and not a swine. Moreover, Anatool had already been baptized and she would not be able to justify it the acceptance of the proposal to her family. Anatool as a human being with a swine's head as transitional type, that she was prepared to accept, but a further compromise she stubbornly rejected. A Christian education was her minimal condition.

'I am not interested in transitional types', the countess said. 'Only remarkable specimens of swine fit into my collection. A swine with a human body can still be classified in this collection. But a human being with a swine's head strikes me as too banal. I respect your will, but I find your obstinacy merely foolish. Christina, you naturally don't know much, I cannot expect that from you. Eminent scholars have proven to me that everything comes down to an adaptation to circumstances. As yet there is still time to determine whether your son shall be a human being or a swine, just as in our own family we determine whether our son shall be an officer in the navy or in the army. All parents are confronted with such problems. Your son can just as well grow to be a human being as to be a swine. It is only a matter of placing him in the right environment. And allow me to say this: the swine have an excellent life with me. None of my sows would want to change places with you, dear Christina. I appreciate your sense of honour, but it would lead you to ruin your son's future. And for this reason you are not a true mother, Christina. A true mother thinks only of the future of her offspring and must therefore sacrifice her own sense of honour. As a human being the future of your son is at the least uncertain. We have an entire staff of servants and maids whose only reason for being there is to ensure the best possible care for my remarkable collection. So as a swine Anatool would have an excellent life with me. Nothing would remind him of his unfortunate descent. So all the benefits are yours. The risk is mine. If Anatool should accidentally lose his swine's head, I would be obliged to remove him from my collection. For me it is a great risk.'

She disappeared, after having embraced Anatool, 'unique et mignon'. The lumberjack saw much truth in her disquisition. 'What', he said, 'swine or human being, all that matters is to make one's way through this world in as good and as decent a fashion as possible. Whether one earns one's bread or one's acorns is not what really counts. And Anatool would have a very good life; he would not have to work for a living. No member of my family has ever come that far yet. We have only had a life of drudgery. From now on Anatool would be the finest among us. Only counts and other great men do not have to work for a living. Anatool would consequently be almost as important as the count himself. He would stand on the highest step of the social ladder, so to speak. Frankly speaking: which one of us could ever be so successful as to become the prize item in the countess's collection? Earning a living without working and being admired by everyone on top of that. I too would have stubbornly resisted if the countess had not explicitly said that she only keeps luxury-swine. Our child's future was made and you have tampered with it.' He bent over little Anatool's cradle and cried: 'Poor son, your future has been spoilt by your mother. Never forget this.'

And turning again to Christina: 'It is better to live a decent life as a swine than a dishonest one as a human being. So say I, Judocus the lumberjack.'

Christina kept her courage. She hoped: the swine's head would certainly disappear. And even if, contrary to all expectations, it were not to disappear, even then, who knows ... surely Anatool's swine's head must have some significance.

Such was the commotion at Anatool's birth.

*Translated from the Dutch by Ortwin de Graef*